Shakespeare's Double Helix

Shakespeare

Series edited by Simon Palfrey and Ewan Fernie

To Be Or Not To Be Douglas Bruster
Shakespeare Thinking Philip Davis
Shakespeare Inside Amy Scott-Douglass
Shakespeare's Modern Collaborators Lukas Erne
Shakespeare and the Political Way Elizabeth Frazer
Godless Shakespeare Eric S. Mallin
Shakespearean Metaphysics Michael Witmore

Henry S. Turner

Shakespeare's Double Helix

Continuum
The Tower Building, 11 York Road, London SE1 7NX
80 Maiden Lane, Suite 704, New York NY 10038

www.continuumbooks.com

© Henry S. Turner 2007

All rights reserved. No part of this publication may be reproduced or transmitted in any form or by any means, electronic or mechanical, including photocopying, recording, or any information storage or retrieval system, without prior permission in writing from the publishers.

Henry S. Turner has asserted his right under the Copyright, Designs and Patents Act, 1988, to be identified as Author of this work.

British Library Cataloguing-in-Publication Data
A catalogue record for this book is available from the British Library.

ISBN: 978-0-8264-9119-0 (hardback)
      978-0-8264-9120-6 (paperback)

Typeset by Kenneth Burnley, Wirral, Cheshire
Printed and bound in Great Britain by MPG Books Ltd, Bodmin, Cornwall

Contents

General Editors' Preface vi
Preface x

When everything seems double 2
That is the true beginning of our end 110
Two intertwining helices 3
Then read the names of the actors; and so grow to a point 111

Index 125

General Editors' Preface

'Shakespeare Now!' represents a new form for new approaches. Whereas academic writing is far too often ascendant and detached, attesting all too clearly to years of specialist training, 'Shakespeare Now!' offers a series of intellectual adventure stories: animate with fresh and often exposed thinking, with ideas still heating in the mind.

This series of 'minigraphs' will thus help to bridge two yawning gaps in current public discourse. First, the gap between scholarly thinking and a public audience: the assumption of academics that they cannot speak to anyone but their peers unless they hopelessly dumb-down their work. Second, the gap between public audience and scholarly thinking: the assumption of regular playgoers, readers, or indeed actors that academics write about the plays at a level of abstraction or specialization that they cannot hope to understand.

But accessibility should not be mistaken for comfort or predictability. Impatience with scholarly obfuscation is usually accompanied by a basic impatience with anything but (supposed) common sense. What this effectively means is a distrust of really thinking, and a disdain for anything that might unsettle conventional assumptions, particularly through crossing or re-drafting formal, political, or theoretical boundaries. We encourage such adventure, and base our claim to a broad audience upon it.

Here, then, is where our series is innovative: no compromising of the sorts of things that can be thought; a commitment to publishing powerful cutting-edge scholarship; *but* a conviction that these things are essentially communicable, that we can find a language that is enterprising, individual, and shareable.

To achieve this we need a form that can capture the genuine challenge and vigor of thinking. Shakespeare is intellectually exciting, and so too are the ideas and debates that thinking about his work can provoke. But published scholarship often fails to communicate much of this. It is difficult to sustain excitement over the 80–120,000 words customary for a monograph: difficult enough for the writer, and perhaps even more so for the reader. Scholarly articles have likewise become a highly formalized mode not only of publication, but also of intellectual production. The brief length of articles means that a concept can be outlined, but its implications or application can rarely be tested in detail. The decline of sustained, exploratory attention to the singularity of a play's language, occasion, or movement is one of the unfortunate results. Often 'the play' is somehow assumed, a known and given thing that is not really worth exploring. So we spend our time pursuing collateral contexts: criticism becomes a belated, historicizing footnote.

Important things have got lost. Above all, any vivid sense as to why we are bothered with these things in the first place. Why read? Why go to plays? Why are they important? How does any pleasure they give relate to any of the things we labor to say about them? In many ways, literary criticism has forgotten affective and political immediacy. It has assumed a shared experience of the plays and then averted the gaze from any such experience, or any testing of it. We want a more ductile and sensitive mode of production; one that has more chance of capturing what people are really thinking and reading about, rather than what the pre-empting imperatives of journal or respectable monographs tend to encourage.

Furthermore, there is a vast world of intellectual possibility – from the past and present – that mainstream Shakespeare criticism has all but ignored. In recent years there has been a move away from 'theory' in literary studies: an aversion to its obscure jargon and complacent self-regard; a sense that its tricks were too easily rehearsed and that the whole game has become one of diminishing returns. This has further encouraged a retreat into the supposed safety of historicism. Of course the best such work is stimulating, revelatory, and indispensable. But too often there is little trace of any struggle; little sense that the writer is coming at the subject afresh, searching for the most appropriate language or method. Alternatively, the prose is so labored that all trace of an urgent story is quite lost.

We want to open up the sorts of thinking – and thinkers – that might help us get at what Shakespeare is doing or why Shakespeare matters. This might include psychology, cognitive science, theology, linguistics, phenomenology, metaphysics, ecology, history, political theory; it can mean other art forms such as music, sculpture, painting, dance; it can mean the critical writing itself becomes a creative act.

In sum, we want the minigraphs to recover what the Renaissance 'essay' form was originally meant to embody. It meant an 'assay' – a trial or a test of something; putting something to the proof; and doing so in a form that is not closed-off and that cannot be reduced to a system. We want to communicate intellectual activity at its most alive: when it is still exciting to the one doing it; when it is questing and open, just as Shakespeare is. Literary criticism – that is, really thinking about words in action, plays as action – can start making a much more creative and vigorous contribution to contemporary intellectual *life*.

SIMON PALFREY AND EWAN FERNIE

Preface

This book reads Shakespeare's *A Midsummer Night's Dream* in order to explore the nature of creativity and experimentation in literature and in scientific research. It considers many problems, among them the nature of theater and *mimesis*, the definition of the 'human', and the value of mixing ideas and substances that are not normally mixed together. A driving impulse behind the book has been to explore the risks and pleasures of genuine research: to engage with that kind of thinking, in any field, that begins by asking questions to which one does not yet know the answers and that releases itself into the unknown. I have approached the process of writing the book in this spirit and have addressed it to a variety of audiences, in the hopes that non-specialists and specialists alike will find it of interest. I have presumed relatively little knowledge about Shakespeare's period or about the intellectual contexts that I think are important for understanding *A Midsummer Night's Dream*. At the same time, I have not tried to simplify the theoretical problems raised by the play or by the scientific arguments I consider, although I have tried to render these problems as clearly as possible. As Francis Crick writes in his memoir *What Mad Pursuit*: 'What makes people really appreciate the connection between two fields is some new and striking result that obviously connects them in a dramatic way. One good example is worth a ton of theoretical arguments. Given that, the bridge between the two fields is soon crowded with research workers eager to join in the new approach.' I like to think that *A Midsummer Night's Dream* provides that good example.

I would also like to thank several friends and colleagues who helped me realize the book: Ewan Fernie and Simon Palfrey, who dreamed up the 'Shakespeare Now!' series, generously invited me to contribute, and provided many helpful comments and provocations; Anna Sandeman, the editor at Continuum, for supporting a book with an unusual format; Linda Garrity, a daily interlocutor in the dialogue of self-creation; Austin Johnson, who gave me a tour of 'science in action' at the stem cell laboratories at UW-Madison; Scott Newstok, who passed on invaluable examples of Shakespeare and biotechnology; David Sedley and the members of the interdisciplinary seminar 'Why We Do Where We Do: Modern Divisions of Science and Literature', sponsored by the John B. Hurford Humanities Center at Haverford College, who invited me to present a draft of the book's right-hand side; Jacques Lezra, with whom I had the pleasure of twice co-teaching a graduate seminar on concepts of life and the human that shaped the book in significant ways; Madhavi Menon, who suggested the book's title; and Rebecca Walkowitz, who inspires me in more ways than I can describe. The Graduate School at the University of Wisconsin-Madison provided research support for the book; through their generosity I had the benefit of two superlative research assistants, Megan Massino and Justin Kolb, both of whom contributed enormously to the materials that I was able to bring together. Any errors and sloppy arguments are mine, but any ingenuity or originality should probably be ascribed to them. I am dedicating this little book to all my students at the University of Wisconsin-Madison, graduate and undergraduate alike, because I have enjoyed teaching them so much and because I think that the promise and the responsibility of new questions lies with people like them.

Readers have two options for reading this book, which consists of two separate but related essays that run parallel to one another on facing pages. Each left-hand page of the book centers on *A Midsummer Night's Dream*, showing how Shakespeare used theater to reflect upon his culture's understanding of human, animal, and artificial life; on problems of hybridity and metamorphosis; on the nature of myth and poetic language; and on occult philosophies of nature. The left-hand essay lingers over the play's language, its characterization, and its use of theatrical resources; it explains the play's structural principles and the complexity of the questions that motivate it. At the same time, each right-hand page of the book advances a series of arguments about *mimesis* in science, about language and naming, about the nature of experiment and how scientific knowledge gets produced, and about how contemporary biotechnology forces us to reconsider our normative definitions of the human and our ideas about life in general. These two essays wrap around the spine of the book itself, like the 'woodbine' and 'the sweet honeysuckle / Gently entwist' that Titania invokes (4.1.41–2) – or like a double helix. Readers may choose to read vertically, following one essay continuously on either the left or the right side and skipping from odd-numbered page to odd-numbered page or from even to even. Alternatively, readers may choose to read horizontally, as they would read any other book, in which case they will digest sections of each facing-page argument simultaneously, the eye wandering across the divide of the page to make spontaneous graftings among ideas and to generate new and entirely unanticipated arguments with each reading and re-reading.

In order to begin the process of producing the proteins necessary to a new cell development, the double helix of DNA

opens;

the

two strands

separate and

then begin to

un wind

HENRY S. TURNER

When everything seems double

Methinks I see these things with parted eye,
When everything seems double. (4.1.188–9)

Shakespeare's Double Helix launches an analysis into the challenges posed by the creative imagination to modern scientific inquiry, and it does so through a reading of *A Midsummer Night's Dream* (c. 1595), a play written when the 'new science' had begun to unsettle the foundations of Western knowledge about the natural world. The play invites us to enter a culture that knew nothing of 'our' humanity: nothing of DNA, nothing of biotechnology, nothing of transplants or of the cell. And yet if we regard *A Midsummer Night's Dream* with the contemporary eye that the 'Shakespeare Now!' series invites, we find startlingly familiar scenes, organized around problems that continue to drive the scientific thought of our own era. Over the course of two essays that twist around shared terms, images, and theoretical problems, *Shakespeare's Double Helix* attempts to restore for modern readers the richness and peculiarity of the 'nature of nature' in Shakespeare's own moment while at the same time showing how the questions he poses provide new resources for understanding the science of the twenty-first century: its languages and forms; its presuppositions, claims, and purposes; its radical power to produce the subjects and objects of a post-human world.

Two intertwining helices

Our model (a joint project of Francis Crick and myself . . . is a strange model and embodies several unusual features. However, since DNA is an unusual substance we are not hesitant in being bold. The main features of the model are (1) the basic structure is helical – it consists of two intertwining helices . . . (2) The helices are not identical but complementary . . . Even if wrong I believe it to be interesting since it provides a concrete example of a structure composed out of complementary chains.

So wrote James Watson on 12 March 1953 to his mentor Max Delbrück, the German theoretical physicist at Cal Tech who had helped to pioneer the field of molecular biology through his analysis of replication in the phage virus. In Berlin, Delbrück's early work had been motivated by an interest in adapting atomic models derived from quantum mechanics to the study of genetic mutation, by which he hoped to solve a paradox fundamental to the evolution of life as we know it. What process permits cells to change suddenly while still remaining stable over time, so that rare mutations are passed on from one generation to the next like a series of perfect copies? How, in short, does the living organism reconcile the accidents of chance with the regularity and order necessary to sustained self-reproduction? A theoretical grasp of the problem, reasoned T. M. Timoféëff-Ressovsky, one of

In recent years critics have become aware of the degree to which Shakespeare's theater was like an unfamiliar mode of technology: to the late-sixteenth-century mind, the theater was best compared to an atlas of maps, to a wonderful museum or encyclopedia, or to the new mechanical globes that suddenly reduced the entire human world into miniature form – and which may have given the name to the Globe Theatre, where Shakespeare's own company performed as of 1599. After the dissolution of the last monasteries but before the building of the first laboratories, Elizabethans thronged to a new machine that could fabricate the truths upon which their culture depended, with equal parts pleasure and profit: to the theater and its 'dramatology.' Because of its novelty, the theater offered poets an unusually free medium in which to explore the many changes in social, religious, intellectual, and political life that were transforming the world as they knew it. If the 'new Philosophy calls all in doubt,' as John Donne famously wrote during a moment of epistemological crisis, the theater only focused its effects, refracting the world of lived experience through a seemingly magical use of objects, language, clothing, and bodies. With a mere word, a learned scholar might conjure a devil and fly on his back to torment the Pope in Rome, a general might summon a thousand troops to fight an Emperor, an entire theater audience might be transported over the seas to France, or to North Africa, or to Asia.

Ever since Plato argued through his teacher Socrates that poets should be banished from his ideal republic, philosophers and statesmen alike have regarded artful language with suspicion because of its traffic in *mimesis*: the imitation, representation, or simulation of life. Poetry was a source of lies and

Delbrück's early German collaborators, might yield stunning laboratory work: through artificial means scientists could 'induce at will certain types or groups of mutation' (452) resulting in categorically new organisms, a process Timoféëff-Ressovsky referred to as 'genetic engineering' (451), in one of the earliest uses of the phrase.

From Berlin in 1934 to Cambridge in 1953 to today's labs across the globe, the notion of genetic engineering remains as controversial as ever; indeed, the specter of a world governed by 100 Hitlers served by an army of supermen is one of the most common objections to cloning and other forms of recombinant DNA technology, along with fears of cancer, loss of individuality, and worry over whether artificially generated entities would have souls. Writing in the *St Louis Post-Dispatch*, the paper's science columnist Dr George Johnson pointed out to his readers that current genetic engineering procedures were not only imprecise and impractical for use on humans but potentially dangerous. The process, he suggests, is best compared to 'inserting a phrase from Shakespeare into a line of code in a computer program': since 'the virus enters the host chromosome just any old place,' Johnson explained, 'and the gene that it inserts itself into is usually damaged, the approach causes mutations, and so almost certainly could cause cancer.'

Whatever Johnson's stylistic peculiarities (we might ask ourselves what mode 'almost certainly could' intends to signify), his offhand analogy is worth considering more closely for several reasons, not the least of which is the opposition it implies between two distinct modes of writing and knowing: the 'literary,' signified by the proper name of Shakespeare, and the 'scientific,' signified by the notion of the computer and its code.

distortions that might corrupt the education of the guardians, Socrates maintained, the men raised to rule over the political community. Those who read poetry or attend the theater learn bad habits, Socrates argued, since 'from the imitation they imbibe the reality,' and this is especially true of watching men who play women's roles on stage, or who imitate slaves, or madmen, or cowards, or craftsmen, or animals and natural processes (395 D–396 E). In order 'to be expert craftsmen of civic liberty, and to pursue nothing else that does not conduce to this,' the guardians should be 'released from all other crafts' (395 C); they should not be like those 'mimics' who dabble in many professions but should, like a good tool, be suited only to one particular purpose, for men can do only one thing well (397 E). Since most men can't even imitate well in both tragedy and comedy, Socrates points out, the noble man will imitate very little; the 'debased' man, however, will not 'shrink from imitating anything and everything. He will think nothing unworthy of himself, so that he will attempt, seriously and in the presence of many, to imitate all things, including . . . claps of thunder, and the noise of wind and hail and axles and pulleys, and the notes of trumpets and flutes and pan-pipes, and the sounds of all instruments, and the cries of dogs, sheep, and birds' (397 A–B). For this reason, Socrates concludes, we must not only regulate the poets but also any craftsman who might represent these kinds of things, lest the guardians be 'bred among symbols of evil, as it were in a pasturage of poisonous herbs' (401 B).

Or in the woods outside of Shakespeare's Athens, where craftsmen clamor to play all parts simultaneously and rehearse plays that mix comedy with tragedy; where faeries mingle with young lovers; where men are transformed into women or into

This opposition in turn implies others: fiction opposed to fact, imagination to reason, word to number, creativity to system, consciousness to automation, sensible flesh to senseless silicon, human to machine. My purpose in this right-hand side of *Shakespeare's Double Helix* is to re-examine these distinctions in order to advance a simple but counter-intuitive argument: that we should regard genetic engineering and biotechnology not simply as a new application of scientific knowledge but rather as a new mode of poetics, and that Shakespeare's own work provides a model for just such an approach.

We may be surprised to find that the distinction between 'literature' and 'science' as two utterly different ways of writing and knowing predates the modern scientific revolution and derives in large measure from the history of *mimesis*, the most important concept in the history of inquiry into drama and theater. This derivation is already visible in the classical period: in his *Poetics*, for instance, Aristotle argues that Empedocles does not 'imitate' when writing of natural philosophy, unlike Homer, who writes fictional 'images' (*eikones*, in Greek). Let us begin, therefore, by considering what the concept of *mimesis* designated in the traditions that Shakespeare inherited and to which we, too, literary critics and scientists alike, are the heirs.

In Book III of his *Republic*, Plato famously used the term *mimesis* to describe the technique of epic poets who recount events by pretending to be someone else: the use of dialogue for fictional characters is 'mimetic,' Plato argues, whereas the telling of events in a poet's own voice (which we would describe as third person narration) is not. *Mimesis* thus implied for Plato a notion of impersonation, whether in words (poetry), in action (drama), or in graphic form (painting), and it especially connoted the

beasts; and where many strange herbs grow. It would be interesting to know exactly what Shakespeare made of Plato's objections, which he undoubtedly knew; only a year before *A Midsummer Night's Dream*, Sir Philip Sidney's famous *Defence of Poesy* had rehearsed many of Plato's arguments and launched a compelling (and charming) counter-attack. Unlike Sidney or his fellow playwrights Thomas Heywood and Ben Jonson, Shakespeare wrote no literary criticism and left no account of his own attitude toward poetics beyond his poems and plays. Ten years later he would go on to write the so-called 'heath' scenes of his masterpiece *King Lear*, in which he threw his title character into the very center of 'claps of thunder, and the noise of wind and hail,' as Socrates might have put it, and used the occasion for one of the most profound meditations on the limits of human will and human power in all of Western literature. In 1595, however, his mood was more lighthearted, and *A Midsummer Night's Dream* offers his most playful response to the antitheatrical arguments of his own moment, which he seems to have regarded with a mixture of impatience and amusement.

Those who objected to the stage – and there were many, from the divines and the moralists of the Church of England to the Lord Mayor of London and other city authorities – felt that it was too powerful to be left unregulated: too real in its simulations, too convincing, and thus too likely to provoke psychological and even physical changes in its audiences. 'Poets in theatres wound the conscience,' complained Stephen Gosson (himself a onetime playwright and former actor) in his *Schoole of Abuse* (1579), 'they arrange comforts of melody, to tickle the ear; costly apparel, to flatter the sight; effeminate gesture, to ravish the sense; and wanton speech, to whet desire to inordinate lust.'

qualities of animation, vivacity, or 'liveness' that these modes of representation could achieve. For this reason, Plato also linked *mimesis* to illusion and deception, an aspect of his thought that has often been emphasized. As Stephen Halliwell has shown, however, Plato's dialogues also use *mimesis* in ways that go beyond notions of realism, reference, or illusion to encompass broader and more neutral ideas of analogy, similarity, and likeness. Music was mimetic because its harmonies reproduced the harmonies of the celestial spheres; artistic creation might represent not simply concrete examples but ideal types, concepts, or generalities, and especially hypothetical states of affairs.

Indeed, for Plato all thinking in general is mimetic to the degree that it makes use of mental images and of word-pictures. In the *Cratylus*, for instance, Plato argues that if we trace our words for things and concepts back far enough, we will reach an originary set of names that cannot be said to derive from any others and that stand in a primary mimetic relationship to the entities that they designate. Much of the *Cratylus* consists of long etymological examples, which Socrates adduces in order to demonstrate the 'natural correctness' of names, and one of the first etymologies argues for natural names on the basis of animal species and generation:

> It is right, I think, to call a lion's offspring a lion and a horse's offspring a horse. I am not speaking of prodigies, such as the birth of some other kind of creature from a horse, but of the natural offspring of each species after its kind. If a horse, contrary to nature, should bring forth a calf, the natural offspring of a cow, it should be called a calf, not a colt, nor if any offspring that is not human should be born from a human

'[London's youth] is greatly corrupted & their manners infected wth many evill & ungodly qualities by reason of the wanton & prophane divises represented on the stages by the sayed players,' wrote the Lord Mayor of London to the Archbishop of Canterbury in 1592, who was no doubt sympathetic. As is well known, the theatrical practice of cross-dressing on the Elizabethan stage, in which boys played all the women's roles, provoked particular outrage. 'The apparel of women . . . is a great provocation of men to lust and leachery,' fumed the Oxford don John Rainolds, 'because a woman's garment being put on a man doeth vehemently touch and move him with the remembrance and imagination of a woman: and the imagination of a thing desirable doth stir up the desire . . . if they do but touch men only with their mouth, they put them to wonderful pain and make them mad: so beautiful boys by kissing do sting and pour secretly in a kind of poison.' The risk, in the eyes of the anti-theatricalists, was a monstrous man-woman produced through theatrical artifice: 'Our apparel was given us as a sign distinctive to discern betwixt sex and sex,' wrote Philip Stubbes, 'and therefore one to wear the apparel of another sex is to participate with the same, and to adulterate the verity of his own kind. Wherefore these women, may not improperly be called Hermaphroditi, that is monsters of both kinds, half women, half men' (1583, F5v).

At the heart of the antitheatrical challenge lay a staggering question: what does it mean to make life, and especially forms of life that depart from a normative category of the 'human' understood to be the measure of all living things? In objecting to the stage, the divines and aldermen were laying a twofold charge, objecting both to the inventive power of a theater that could simulate life so convincingly, 'denaturing' the world through

being, should that other offspring be called a human being; and the same applies to trees and all the rest. Do you not agree? (393 C)

We will not be surprised to find that Hermogenes, Socrates's interlocutor, does agree, since the philosophy of language that Socrates offers depends on natural or essential distinctions among living kinds and between human living beings and other kinds of living beings, so that mixed species are either clarified linguistically, through the orthopedics of the name (calf rather than colt), or they remain outside language altogether and thus outside the categories of nature, reality, and being.

In order to explain the primary concepts or names of philosophy, Socrates then advances a version of a materialist argument that he elsewhere rejects, suggesting that since 'the nature of things really is such that nothing is at rest or stable, but everything is flowing and moving and always full of constant motion and generation,' so in these cases 'the names . . . are given under the assumption that the things named are moving and flowing and being generated' (411 C). A series of further etymologies follow, and the dialogue seems to have excavated a primal level of things emerging into names, including things that are not really things but rather processes or energies or cascades of particles. These primary elements must also be a kind of name, since 'there is but one principle of correctness in all names, the earliest as well as the latest,' and the 'correctness of all the names we have discussed was based upon the intention of showing the nature of the things named' (422 D).

Socrates would like to imagine the elements of nature as existing prior to these primary names, but as the dialogue proceeds

imitations and unrestricted language, and to the fact that these creatures who stalked the stage – these artificial heroes, these kings of gesture, these gods of shadow, these ghosts, spirits, clowns, and other parodies of men – might at any moment flagrantly disregard the boundaries that seemed to secure a civilized humanity. Beggar a father, prostitute a sister, murder a monarch, mock a bishop, fool a judge; charm, seduce, fondle, force; chase bestial pleasures and consume all things; swear false oaths, blaspheme, and rail against authority; turn Fortune to self-advantage and a deaf ear to mothers' pleading; bake children into meat-pies; consort with witches and conjure demons; curse the rain, the sun, the moon, the stars, the elements; speak nonsense; play the fool; become an ass – all these, and more, the poets bodied forth with relish upon their stages, and Shakespeare first among them.

Confronted by the suspicion of authorities toward his professional medium, Shakespeare began to explore the properties of theatrical performance more closely and soon found himself immersed in a poetic laboratory, one in which the poet might exercise his power to create fanciful substances and inhuman forces as freely as possible. And the Roman sources that informed his work on *A Midsummer Night's Dream* – Ovidian myth and that most outrageous and delightful novel of antiquity, Apuleius's *The Golden Ass*, in which the narrator Lucius is transformed into an ass by an act of magic and undergoes a series of misadventures – had forced into the foreground a range of questions that lay at the heart of early modern critiques of the stage and that preoccupied some of the most adventurous thinkers of Shakespeare's generation. What forces account for physical change? How far can natural objects be altered and still

he seems equally to think that these elements emerge only in the moment of naming, and he describes this process using several analogies, each of which is worth examining more closely. One model for this primary language is picture-making, in which the name is like a diagram, overlaying a reality that exists independently of and outside of language. 'A name,' Socrates concludes, 'is a vocal imitation of that which is imitated, and he who imitates with his voice names that which he imitates' (423 B), but this does not include bleating like a sheep or crowing like a cock, Puck-like, which is a kind of 'musical' imitation but not naming. Naming is specifically the imitation of the 'essential nature of each thing by means of letters and syllables' (423 E), and he who does so correctly is the 'name-maker,' the *onomastikos*, and the process he employs we could call an onomapoietics, applying letters and syllables to things 'just as painters, when they wish to produce an imitation, sometimes use only red . . . and sometimes mix many colours . . . Just as in our comparison we made the picture by the art of painting, so now we shall make language by the art of naming, or rhetoric, or whatever it be' (424 C–425 B).

Socrates also compares naming to dramatic action and to miming in a famous passage:

> If we had no voice or tongue and wished to make things clear to one another, should we not try, as dumb people actually do, to make signs with our hands and head and person generally? . . . If we wished to designate that which is above and is light, we should, I fancy, raise our hand towards heaven in imitation of the nature of the thing in question; but if the things to be designated were below or heavy, we should extend our hands towards the ground; and if we wished to mention a galloping

remain natural? How are natural processes to be distinguished from the power of magic, the stars, or the machine? How are occult processes to be understood if they remain by definition invisible and immaterial? What distinguishes God from man, man from woman, human from animal, beast from plant, plant from stone? In a world governed by ecclesiastical policy and absolute law, what place remains for the hybrid? For imagination and artifice? For metamorphosis and desire? For dream and myth?

Like many playwrights writing for the public theaters, Shakespeare found in classical myth and romance an inexhaustible fund for a new commercial drama that could meet the demands of a paying audience. But classical sources were notoriously problematic for a playwright, since even with careful handling they might introduce unintended resonances into a scene or suggest unflattering analogies to contemporary circumstances. This interpretive volatility resulted from the primary unit of mythic discourse: the 'symbol,' as the classicist Jean-Pierre Vernant has eloquently described it:

> [the symbol] does not belong to the order of intellectual comprehension, as the sign does, but rather to that of affectivity and desire . . . In contrast to the sign, ideally univocal, the symbol is polysemic . . . Signs and categories of signs can be defined precisely; they each have a distinctive function; they are employed in regular combinations. In contrast, symbols possess a fluidity and freedom that enable them to shift from one form to another and to amalgamate the most diverse domains within one dynamic structure. They can efface the boundaries that normally separate the different sectors of

horse or any other animal, we should, of course, make our bodily attitudes as much like theirs as possible . . . For the expression of anything, I fancy, would be accomplished by bodily imitation of that which was to be expressed. (422 E–423 B)

Since 'speaking is an action' and 'naming is a part of speaking,' then 'naming [is] also a kind of action, if speaking is a kind of action concerned with things' (387 C).

This association between language and action is one of the most interesting aspects of the *Cratylus*, and the word for 'action' that Socrates uses is the term *praxis*, a fundamental term in classical ethical philosophy as well as in Aristotle's later *Poetics*, where *praxis* becomes both the form and the substance of drama, its mode of representation as well as the content that it represents. But Socrates uses *praxis* to refer to actions that modify a natural object in an artful or technological way, to use a term more closely related to *techne*, the Greek for 'art.' Every action of this type, Socrates argues, has a natural instrument or proper method, just as weaving and boring require a shuttle and a borer. The weaver who weaves well uses a shuttle that has been formed by the carpenter, and the carpenter is not any person but 'he who has the skill' (*techne*). The hole-maker uses the borer, which has been fashioned by the work of the smith, and the smith is likewise not any man but 'he who has the skill.' This skill lies in discerning the nature of the action and the nature of the material and then making a tool that is 'naturally fitted for each purpose', according to the 'form' (*eidos*) that is proper to it and by which we know it to be an 'absolute or real' tool (389 B–C). If when we weave, we 'separate the mingled threads of warp and woof,' Socrates

> reality and convey in the reflection of a network of mutual relationships the reciprocal effects and the interpenetration of human and social factors, natural forces, and supernatural Powers (237–8) . . . Myth is not only characterized by its polysemy and by the interlocking of its many different codes. In the unfolding of its narrative and the selection of the semantic fields it uses, it brings into play shifts, slides, tensions, and oscillations between the very terms that are distinguished or opposed in its categorical framework; it is as if, while being mutually exclusive these terms at the same time in some way imply one another. Thus myth brings into operation a form of logic that we may describe, in contrast to the logic of non-contradiction of the philosophers, as a logic of the ambiguous, the equivocal, a logic of polarity. (260)

The particular purpose of myth, Vernant argues, was to provide an alternative language for considering matters that conventional philosophy failed to comprehend. What thing lurks at the limits of apprehension, thrusting itself forward toward the light of understanding only to turn away or transform itself into something our reason strains to grasp? Monstrosities and pollutions, the restless movement of pure force, fears and passions that engulf the blinking eye of selfhood: all were unspeakable except through the peculiar capacities of mythic language and its scandalous ability to furnish ever more compelling scenes. The philosophers themselves, meanwhile, could respond in one of only two defensive gestures: to swallow the symbol or to exclude it, to digest or to expel, to translate the code or to interrupt it:

argues, then the name, like the shuttle, is an 'instrument of teaching and of separating reality.' As an instrument, the name either weaves or perforates, depending on how you handle it; it either knits reality together or punctures holes in the primary flux of the world and remains as a nominal leftover or a kind of hanging chad. Under the cover of a dialogue about the origin of language, we have found a philosophy of action and of acting, including in a theatrical sense, and I will indulge in an act of onomapoietics and call this philosophy of language-as-action and action-as-language a 'dramatology.' And this dramatology has turned out not simply to be a philosophy of mere 'imitation' but a philosophy of transforming the world by acting upon it in a technological way, using words as an *organon* or tool in the drama of coming to know the world.

What's in a name?

Shakespeare himself takes two contrasting positions on the problem of naming. In a famous passage toward the end of *A Midsummer Night's Dream*, Theseus adopts a dramatological argument very similar to that of Socrates when he argues that a poet's power stems from the fact that he wields the name as a tool to create something out of nothing:

> The poet's eye, in a fine frenzy rolling,
> Doth glance from heaven to earth, from earth to heaven,
> And as imagination bodies forth,
> The forms of things unknown, the poet's pen
> Turns them into shapes, and gives to airy nothing
> A local habitation and a name.
>
> (5.1.12–17)

The ancient Greeks themselves did not simply relegate myth, in the name of *logos*, to the shadows of unreason and the untruths of fiction. They continued to make literary use of it as the common treasure-house on which their culture could draw in order to remain alive and perpetuate itself. Furthermore, as early as the archaic period they recognized its value as a means of teaching, but an obscure and secret one. They considered it to have some function as truth, but this truth could not be formulated directly, and before it could be grasped had to be translated into another language for which the narrative text was only an allegorical expression . . . Myth was thus purged of its absurdities, implausibilities, and immorality, all of which scandalized reason. But this was achieved only at the cost of jettisoning myth's own fundamental character, refusing to take it literally and making it say something quite different from what it actually told . . . Plato often appears to reject *muthos* utterly as, for instance, when in the *Philebus* (14a) he writes of an argument, *logos*, which, being undermined by its own internal contradictions, destroys itself as if it were a *muthos*; or when, in the *Phaedo* (61b), he has Socrates say that *muthos* is not his affair but that of the poets – those same poets who, in the *Republic*, are exiled, as liars, from the city. (220–1)

But despite his hostility to *muthos*, even Plato:

grants an important place in his writing to myth, as a means of expressing both those things that lie beyond and those that fall short of strictly philosophical language . . . how would it be possible to speak philosophically of Becoming, which, in

Nothing truly new can ever emerge or ever be known to us, Theseus suggests – no genuine 'discovery' of any sort can take place – without this moment of mimetic 'naming' that produces a thing and then allows us to consider it, handle it, test it, theorize about it, or even ban and forbid it. Like Socrates, Theseus seems to believe in the power of language to birth the world, and like Socrates he would jealously preserve authority over naming for the lawgiver and exclude the poets from Athens lest they assume this dangerous power for themselves. In Theseus's speech, the world does not pre-exist our words or the concepts that words allow us to think, and to use language is inevitably to brush up against the alien, the heterodox, and the radically new – an encounter best restricted, in his view, to the judge and ruler. But in *Romeo and Juliet*, Shakespeare takes the opposite position:

> O Romeo, Romeo! wherefore art thou Romeo?
> Deny thy father and refuse thy name;
> . . .
> 'Tis but thy name that is my enemy;
> Thou art thyself, though not a Montague.
> What's Montague? It is nor hand, nor foot,
> Nor arm nor face, nor any other part
> Belonging to a man. O, be some other name!
> What's in a name? That which we call a rose
> By any other name would smell as sweet;
> So Romeo would, were he not Romeo call'd,
> Retain that dear perfection which he owes
> Without that title. Romeo, doff thy name,
> And for that name which is no part of thee
> Take all myself.
>
> (2.1.76–92)

its constant change, is subject to the blind causality of necessity? Becoming is too much a part of the irrational for any rigorous argument to be applied to it . . . Thus, in referring to the gods or the birth of the world, it is impossible to use *logoi homologoumenoi*, totally coherent arguments. One must make do with a plausible fable, *eikota muthon* (*Timaeus*, 29b and c). (221)

If in the *Republic* Socrates dismisses poets and players, in many other dialogues he himself behaves as a kind of poet, using invented analogies and outlandish stories to persuade his interlocutors of arguments that they might never accept at face value. This paradox – that one of the most ancient critiques of poetry and drama was itself delivered in dialogue form by speakers who resembled dramatic characters and who resorted to poetic expression – was not lost on Renaissance writers, as no less a figure than Sidney testifies: 'And truly even Plato,' he pointed out in his *Defence*, 'whosoever well considereth shall find that in the body of his work, though the inside and strength were Philosophy, the skin, as it were, and beauty depended most of Poetry: for all standeth upon dialogues, wherein he feigneth many honest burgesses of Athens to speak of such matters, that, if they had been set on the rack, they would never have confessed them' (97.10–16). Most Renaissance playwrights took a safe course and adapted their mythic sources, domesticating them into genre vehicles that were sure to please the widest possible audience, no matter how popular or how powerful. A few, however, notably Christopher Marlowe and Shakespeare himself, managed to tap the heterodox potential that had long resided in mythic discourse in

Juliet would like very much to separate the man she desires from the name that represents him, much the way she can separate the scent of the rose from the name that classifies it. She craves a kind of radical anonymity for her lover, one that negates social and legal identity and leaves only surfaces, odors, and flesh, as her desire produces a blazon that disaggregates 'Romeo' into his component parts while coyly refusing to name the part that most distinguishes him *as* a man – what importance could a mere *name* have, she seems to ask, when compared to these more important and more desirable features? And since the term 'name' was a more general one in Shakespeare's period for the part of speech we call a 'noun,' we can be sure that Juliet is considering a philosophical question as well as an affair of the heart.

Most scientists take a philosophical position similar to Juliet's, and they believe that the legitimacy of their entire enterprise depends upon it (which is why they often take a dim view of literary theorists). Science, in their view, is the antidote to the mythic power of the name. But is the power of the molecular biologist not a kind of *onomapoetics*? Does she not manipulate names as much as she handles specimens? Researchers in the Department of Molecular and Cell Biology at Berkeley and the Mount Sinai School of Medicine have recently identified two new fibroblast growth factor genes (so-called FGF genes), named 'Thisbe' (ths) and 'Pyramus' (pyr), in *Drosophilia*, the common fruitfly and the model species of modern genetics. These genes, which indirectly code for the proteins necessary to the development of cardiac tissue, are always found linked closely together in sequence, and since the mutant phenotype (a fly without either ths or pyr) exhibits incomplete cardiac development, the authors named their discovery 'for the "heartbroken" lovers described in

order to give dramatic form to the unspeakable, the irrational, and the radically new.

Shakespeare's mythic theater

Audiences who return to read *A Midsummer Night's Dream* can be surprised by some of its darker shadows, which stretch longer across the page than across the stages of most summer festivals. As in all myths, desire is the life-blood of its action, the underlying motivation of every character and a force that becomes by turns romantic, giddy, obsessive, violent, and supernatural in its power. At the opening of the play, Theseus, the son of Aethra and Poseidon, and thus like many mythological heroes the hybrid product of a union between a human and an immortal, waits impatiently to consummate his marriage with Hippolyta, the Amazon queen he has abducted as a spoil of battle. By immediately introducing the marriage theme, Shakespeare announces a shift from the world of epic – the world of Chaucer's *Knight's Tale*, along with Ovid and Apuleius one of the most important sources for his play – to that of comedy, as Theseus himself insists:

> Go, Philostrate,
> Stir up the Athenian youth to merriments
> Awake the pert and nimble spirit of mirth.
> Turn melancholy forth to funerals –
> The pale companion is not for our pomp.
> (1.1.11–15)

Ovid's *Metamorphoses*,' the very myth of loss and transformation that inspired Shakespeare to write the plot of the 'rude mechanicals' in *A Midsummer Night's Dream* and which Bottom and company submit to such monstrous 'disfigurement' at Theseus's court. Peter Quince is like a misfiring regulator-gene whose breakdown produces a garbled script and a wild theatrical hybrid: 'A tedious brief scene of young Pyramus / And his love Thisbe: very tragical mirth' (5.1.56–7).

And perhaps there is more 'science' in Shakespeare than we realize. Announcing a 2002 breakthrough in the genetics of fragrance, Nancy Ekardt, the editor of *The Plant Cell*, the journal of the American Society of Plant Biologists, reported that 'A genomic approach that encompasses cDNA sequencing, microarray gene expression analysis, chemical analysis of the volatile composition of rose petals, and biochemical analysis of candidate proteins' has given researchers at the Institute of Plant Sciences and Genetics in Agriculture at Hebrew University of Jerusalem and the Department of Molecular, Cellular, and Developmental Biology at the University of Michigan new insight into one of Shakespeare's most famous images:

'What's in a name? That which we call a rose, by any other word [sic] would smell as sweet.' So declares Juliet as she laments the name of her beloved in Shakespeare's *Romeo and Juliet*. The fact is, today there are numerous varieties of ornamental rose that produce little or no fragrance . . . Floral scents are complex mixtures of chemicals. Although there are several main groups of compounds (e.g., monoterpenes and sesquiterpenes, aromatic alcohols, and esters), floral scent is highly species specific, and almost no two species produce

But the shadow of violence lingers offstage despite Theseus's cheerful invocation: in his command to 'stir up the youth,' echoing charges of rebellion and the rough 'merriments' of London apprentices that disrupted the streets of sixteenth-century London, including mob assaults on foreigners and the destruction of property; or in the promise to Hippolyta that immediately follows:

> Hippolyta, I wooed thee with my sword,
> And won thy love doing thee injuries.
> But I will wed thee in another key –
> With pomp, with triumph, and with reveling.
> (1.1.16–19)

In four lines Shakespeare deftly sketches a portrait of remarkable psychological complexity: Theseus offers an oblique apology, muffles it with a wishful distortion of past action (not violence but love; not wounds but wooing), and then punctuates it with a lavish, compensatory gesture that celebrates his own power and righteousness. The speech has a whiff of both sadism and megalomania, and it is easy to imagine that Theseus is 'over-full of self-affairs,' as he declares moments later (1.1.113). Forced to delay his gratification – the moon 'lingers my desires' (1.1.4) – he can only hope that 'Four days will quickly steep themselves in night / Four nights will quickly dream away the time' (1.1.6–7), as Hippolyta reassures him, perhaps fondly, perhaps with a note of grim determination. These rotations of day to night mark the putative time of the action that ensues, although like all dreams *A Midsummer Night's Dream* compresses events so that they seem to occur over a much shorter period; the play's

identical mixtures of scent compounds. Even within species, often there is a great deal of variability in scent production. Rose is a prime example: many cultivars produce little or no scent, and among those that do, there is considerable variability in the type of scent produced.

The difficulty and relevance of the research in the rose study stems (as it were) from the 'invisibility' and 'dynamic nature' of floral fragrance, and the authors of the study justify their research by pointing out its economic implications: 'For centuries, rose has been the most important crop in the floriculture industry,' they write:

> The genus *Rosa* includes 200 species and [more than] 18,000 cultivars . . . At an annual value of [approximately] $10 billion, roses are used as cut flowers, potted plants, and garden plants. Their economic importance also lies in the use of their petals as a source of natural fragrances and flavorings. The damask rose (*Rosa damascena*) is the most important species used to produce rose water, attar of rose, and essential oils in the perfumery industry.

The 'damask rose' was also, of course, one of the most important species in the *poetry* industry of the late sixteenth century, a ubiquitous trope that Shakespeare employed in what is arguably his most famous sonnet, number 130 ('My mistress' eyes are nothing like the sun'):

> I have seen roses damasked, red and white,
> But no such roses see I in her cheeks
>
> (5–6)

scenes, Shakespeare seems to suggest, unfold like one of Theseus's own dreams, an inversion and a displacement of an urgent desire for Hippolyta that has been impeded by the requirements of civil law.

This 'mythic' conflict between an absolute, arbitrary law that remains removed from embodied persons and the driving force of a fully embodied desire that is entirely beyond reason and its categories seems to have held a particular fascination for Shakespeare, since it motivates many of his plays, above all *Romeo and Juliet* – written in the same moment as *A Midsummer Night's Dream* and which he based on the same myth of Pyramus and Thisbe that Bottom and the 'rude mechanicals' rehearse in the forest outside Athens – and *Measure for Measure*. As the first scene shifts to Theseus's court, Shakespeare transposes his theme onto the axis of filial relationship: when Hermia declares that she loves Lysander and not Demetrius, her father Egeus's choice of suitor, Egeus invokes paternal authority as an absolute right of property, 'to dispose of her / Which shall be either to this gentleman / Or to her death, according to our law / Immediately provided in that case' (1.1.42–5). In an attempt to resolve the conflict between father and daughter (a conflict between law and desire not unlike his own), Theseus immediately confirms Egeus's claim by arguing that his power over his daughter's life resides in a principle of divine paternity that any actual law would simply reinforce or cast in more abstract, less personal form:

> look you arm yourself
> To fit your fancies to your father's will,
> Or else the law of Athens yields you up –

Through a supplemental twist of sheer ingenuity, the image of the 'damask rose' becomes a metaphor for the poetic use of metaphor itself: Shakespeare's point is that poetic cliché can never capture the living particularity of a lover, and he argues here, like Juliet, for a kind of literalism or realism rooted in perception rather than in language (all while using poetry and figurative language, of course, to do so). The methods employed by the authors of the rose study may prove useful to literary critics, since, in the words of Ekardt, they:

> can be expected to yield more valuable information about the genetics and biochemistry of floral scent production. In the meantime, we may be content to contemplate which fragrance-related genes inspired Shakespeare to pen some of his most exquisite lines:
>
>> I know a bank where the wild thyme blows,
>> Where oxlips and the nodding violet grows,
>> Quite over-canopied with luscious woodbine,
>> With sweet musk-roses and with eglantine:
>> There sleeps Titania sometime of the night,
>> Lull'd in these flowers with dances and delight

Ekardt seems unaware of a recent breakthrough: some six months earlier, an unlikely creative collaboration between the Royal Shakespeare Company and the Royal Society of Chemistry had resulted in a new perfume, 'Puck's Potion,' based on the fragrances tangerine, bergamot, white pepper, cloves, and violets. 'There are scores of references to plants and herbs in Shakespeare, who was obviously very knowledgeable about their

> Which by no means we may extenuate –
> To death or to a vow of single life.
> (1.1.117–21)

In calmly offering two absolute alternatives, virginity or death, Theseus frames an opposition in which either choice evacuates desire into the airless, absolute space of a non-human legal principle.

In this way the first scene functions as an initial unit in an unfolding theatrical series, such that the composition of the play as a whole becomes kaleidoscopic or crystalline, each scene taking shape around a set of thematic, verbal, or conceptual relationships that soon modulate and give way to others. Here, for instance, a primary opposition between law and desire, or between public ceremony and private satisfaction, or between mirth and melancholy, or between sex and virginity, or between life and death – for which pair can be said to be truly 'primary'? – generates others in neighboring but different registers: the opposition between sun and moon (a cosmological image), between waking and sleeping (a psychological and physiological one), or between solipsism and sympathy (a relation that is at once psychological, emotional, and ethical). Although the script and the actor's technique may lend to any one pair a momentary emphasis, the scene has been woven out of many strands that coalesce and then disintegrate before the audience, combining and recombining to generate, gradually and in fragments, the 'play' that comes to fill the stage.

The play's kaleidoscopic and dreamlike quality is only heightened as we enter the forest and encounter Oberon and Titania,

real and mythical potency,' Charles Sell, head of organic chemistry at Quest International, the fragrance company commissioned to produce the 'Potion,' was quoted in the press at the time. These included the wild violet species 'Heart's Ease,' named for its palliative properties in cases of asthma, epilepsy, bronchitis and some heart ailments. Appropriately, the perfume had been designed with Valentine's Day in mind, although Claire McLoughlin of the Royal Society of Chemistry said that it would not be released publicly (perhaps for obvious reasons). Members of the company were reportedly testing it on their Titania.

Stars, genes, and other particles in need of names

No doubt many scientists would regard references to Shakespeare in a lab report with a mixture of amusement and skepticism, feeling reflexively that the poetry provides imaginative color but little of scientific substance. 'All this is entertaining enough,' they say, 'but what's the real point?' In Shakespeare's own period, Bacon and Galileo offered some of the most famous statements of what was then a newly 'scientific' position, one that distinguished the work of the scientist from that of the poet in part by separating words from the things they seemed to designate. Bacon argued in his *Advancement of Learning* (1605) that 'the first distemper of learning' is 'when men study words and not matter,' and therefore 'Pygmalion's frenzy is a good emblem or portraiture of this vanity: for words are but images of matter; and except they have life of reason and invention, to fall in love with them is all one as to fall in love with a picture' (26). The fate of Galileo is even more instructive, since it suggests how much could be at stake in debates over language, things, and 'naming.'

both of whom were almost certainly doubled in performance by the same actors who played Theseus and Hippolyta, a stage practice that often survives today. As in all myths (and in all dreams), the action unfolds through a series of displacements, condensations, and reversals of image: the sun sets and the moon rises, becoming one of the most significant recurring symbols of the play; the public, political, and lawful community of Athens gives way to the 'nightrule' of the woods, with its shifting appearances and superhuman conflicts. Both Oberon and Titania operate with a planetary reach: 'We the globe can compass soon, / Swifter than the wand'ring moon' (4.1.96–7), Oberon reminds Puck, who promises to 'put a girdle round about the earth / In forty minutes' as he exits to retrieve the mysterious flower that Oberon will use to torment Titania ('Shakespeare was not far wrong,' the critic Jan Kott observed in 1964, 'the first Russian sputnik encircled the earth in forty-seven minutes.').

The proximate cause of the quarrel between Oberon and Titania is an Indian boy whom Titania has adopted after the death of his mother and whom Oberon now desires for his own reasons, which remain obscure: 'jealous Oberon would have the child / Knight of his train, to trace the forests wild' (2.1.24–5), Puck declares, but his motivations seem equally to include a bid for dominance, willful competitiveness, and, as Kott first suggested, erotic interest. Titania refuses, reproaching him for jealousy and faithlessness ('in the shape of Corin . . . Playing on pipes of corn, and versing love / To amorous Phillida' [2.1.66–8]); and Oberon responds in kind, accusing her of 'love to Theseus' and of assisting the mythic hero in his sexual conquests:

In his letters of 1612 debating whether sunspots were actually stars, Galileo argued that:

> It is indeed true that I am quibbling over names, while I know that anyone may impose them to suit himself. So long as a man does not think that by names he can confer inherent and essential properties on things, it would make little difference whether he calls these [phenomena] 'stars.' Thus the novae of 1572 and 1604 were called 'stars,' and meteorologists call comets and meteors 'stars,' and for that matter lovers and poets so refer to the eyes of their ladyloves:
>
>> When Astolfo's successor is seen
>> By the glance of those two smiling stars.
>> [citing *Orlando Furioso* VII, 27, 1–2] (139)

With typical sarcasm, Galileo invokes the romance poet in order to show how promiscuous language can be, so that he may better assert an ontology of astronomical bodies beyond the conventions of human language. In his later debate with the Jesuit Horatio Grassi over the nature of the comets that had appeared in the skies of Europe in the fall of 1618, Galileo again distinguishes between the world as it has been described by the poets and the world as it truly exists:

> In Sarsi [Grassi's pseudonym] I seem to discern the firm belief that in philosophizing one must support oneself upon the opinion of some celebrated author, as if our minds ought to remain completely sterile and barren unless wedded to the reasoning of some other person. Possibly he thinks that

> Didst not thou lead him through the glimmering night
> From Perigouna whom he ravished,
> And make him with fair Aegles break his faith,
> With Ariadne and Antiopa?
>
> (2.1.77–80)

But Titania denies the charge, dismissing it again as 'the forgeries of jealousy' (2.1.81), the third time in 60 lines that Shakespeare ascribes the emotion to Oberon. The effect heightens our perception of the character as a negative avatar of Theseus, one who embodies love's distorting, possessive impulses and whose exaggerated sense of injustice prompts him to cruelty and gratuitous acts of humiliation: 'Thou shalt not from this grove / Till I torment thee for this injury' (2.1.146–7), he mutters, using words identical to those spoken by Theseus to Hippolyta at the opening of the play, and later vows to 'make her full of hateful fantasies' (2.1.258). The entire exchange opens a glimpse onto a metaphysical dimension to Theseus's character that remains latent at the opening of the play: if the scenes in Athens establish his role as ruler over men and as a legendary but mortal hero, the forest scenes make evident his divinity and ontological reach – he moves easily among both human and superhuman beings, making love with the capriciousness and ignorance of consequence that characterizes any true god.

Nature's distemperature

Oberon and Titania are mythic figures who resemble ourselves – they have all our passion, all our irrational demands,

philosophy is a book of fiction by some writer, like the *Iliad* or *Orlando Furioso*, productions in which the least important thing is whether what is written there is true. Well, Sarsi, that is not how matters stand. Philosophy is written in this grand book, the universe, which stands continually open to our gaze. But the book cannot be understood unless one first learns to comprehend the language and read the letters in which it is composed. It is written in the language of mathematics, and its characters are triangles, circles, and other geometric figures without which it is humanly impossible to understand a single word of it; without these, one wanders about in a dark labyrinth . . . [Sarsi] seems not to know that fables and fictions are in a way essential to poetry, which could not exist without them, while any sort of falsehood is so abhorrent to nature that it is as absent there as darkness in light. (238)

By distinguishing words from things, mathematics from *mimesis*, and science from poetry, Galileo is in fact clearing a space for his own authority to name the world; he invests argumentative energy in the problem of naming because the authority of the man of science resides to a large degree in his capacity to apply names correctly, language being one of many tools and instruments, along with telescopes, lenses, thought-experiments, and geometrical diagrams, that he uses to compose a new and newly authoritative scientific method. And in the early seventeenth century the opposition between poetry and science stands in for another far more dangerous opposition between the book of Nature and the book of Scripture: to end up on the wrong side of *this* distinction, as Galileo soon realized, was to end up in the hands of the Inquisition.

and all our desires – but they also serve Shakespeare as larger than life personifications of natural principles, as vehicles for colossal inhuman forces that only a mythic theater could comprehend. For what, after all, is 'nature,' Shakespeare seems to ask, especially when it so often reveals itself in ways that seem totally alien to us? Does 'nature' include the 'human mortals,' as Titania calls us, or does it always exceed us, and even show us to be less 'human' than we imagine ourselves to be? Is 'nature' a place, like a wood, or river, or sea, or glen? Or is it a system of relationships: a series of causes, effects, and transformations that are difficult to see or locate in any tangible way? In one of the play's most famous passages, Shakespeare uses mythic symbols to describe the 'complexity' of natural forces, in the sense that modern science gives the term – the way in which many local factors quickly combine to produce effects that are impossible to anticipate and very difficult to model – with a clarity that would astonish a modern ecologist:

> Therefore the winds, piping to us in vain,
> As in revenge have sucked up from the sea
> Contagious fogs which, falling in the land,
> Hath every pelting river made so proud
> That they have overborne their continents.
> The ox hath therefore stretched his yoke in vain,
> The ploughman lost his sweat, and the green corn
> Hath rotted ere his youth attained a beard.
> The fold stands empty in the drowned field,
> And crows are fatted with the murrain flock.
> The nine men's morris is filled up with mud,

The arguments of Bacon and Galileo have contributed much to our conventional notions about literature and science, and with these to our sense about what it means to be 'modern,' although like the sociologist of science Bruno Latour, I do not think we have ever been as 'modern' as we imagine – nor have we ever been as 'human.' But it is important to recognize that any sharp distinction between literary and scientific modes of representation relies on premises that approach tautology. If poetry is 'mimetic' because it makes fictions, and 'science' is not mimetic because it deals with facts, and facts are real, true, measurable, and empirical – i.e. not mimetic – then we are led to conclude that science is not mimetic because it specializes in non-mimetic modes of expression, a definition that presumes what it tries to explain. We can short-circuit these self-reproducing tautologies, however, simply by cutting and splicing them: why not regard science, and especially contemporary genetic engineering and biotechnology, as a new chapter in the long history of *mimesis* and poetics?

It should come as no surprise, in the first place, that modern science cannot function without a concept of *mimesis*: the debate over 'art and nature' that originates with Plato and Aristotle and which so fascinated Shakespeare and his contemporaries had always been framed as a problem of imitation. Many sixteenth-century writers invoked the principle that 'art imitates nature' in order to explain any concept of skill, habit, technique, rule, or norm, not simply in poetry or the occult sciences but in grammar, logic, and rhetoric, in theories of education and governance and law. But the principle was especially important to accounts of the mechanical arts, the fields of practical and applied knowledge rather than of theory, of craft and artifice or

> And the quaint mazes in the wanton green
> For lack of tread are undistinguishable.
>
> (2.1.88–100)

Many of the phenomena that bedevil the scientist confronted by ecocides and global warming can be glimpsed here, as Shakespeare tries to fit the terrifying unpredictability, the hugeness and impersonality of forces that operate randomly and with no regard for human interests into a logical sequence that stems from an identifiable cause. The passage alternates between assigning the animating force of nature to a series of interlinked subjective agents – the winds, the fogs, the rivers, the moon, the seasons, back to the quarrel itself, all yoked throughout the passage by a series of logical 'therefores' – and juxtaposing states of affairs that exist independently without clear cause or motivation, as marked by the use of the passive voice and the past present tense. In the process the logical sequence begins to drift; by the middle of the passage, 'therefore' operates without a clear referent, and it becomes unclear precisely *why* the ox stretches his yoke in vain, or the ploughman toils uselessly, or the corn shrivels on the stalk; why the fold stands empty, why water stagnates where it should not, why crows suddenly alight to feed on diseased carcasses – we have moved beyond generic images of cosmic imbalance to something much more sinister, a grotesque but nevertheless still entirely 'natural' order that lives immanently and in potential, behind or even inside a nature that we human mortals can recognize only as a 'quaint maze' or as a locale for a festive holiday.

This 'nature' is, for a moment, revealed as entirely a-theistic, beyond consolation, redemptive vision, or design:

of what we would call 'technology.' As early as the twelfth century, Hugh of St Victor had argued that 'the products of artificers, while not nature, imitate nature, and in the design by which they imitate, they express the form of their exemplar, which is nature' (Newman, 424). 'One sort of art perfects that which nature cannot complete,' Aristotle wrote in his *Physics* (2.8 199a), 'while another sort imitates nature' (Newman, 428); for this reason, the mechanical arts were a 'mixed' form of knowledge that scholastic writers called 'adulterate.' 'The human work,' Hugh writes, 'because it is not nature but only imitative of nature, is fitly called mechanical, that is adulterate' (Newman, 424).

An expanded sense of *mimesis* not as mere copying or imitation but as a dynamic process of invention and creation becomes prominent in Aristotle's *Poetics*, where it is subsumed under his more general notion of *poiesis*. In classical Greek, *poiesis* designated the act of material as well as of verbal making, including shoes and ships and houses; the modern term 'fiction,' another term closely related to *mimesis* and derived from the Latin *fingere*, to fashion or to form, retains a distant resonance of these ideas. If the shoemaker makes shoes and the builder makes buildings, Aristotle had reasoned, then the poet makes fictions or plots: patterns of action that simulate the actions of human life but which the poet artificially structures in order to produce a variety of intellectual and emotional effects in his readers. In the theater, this process of mimetic 'making' literally becomes a dramatology: a process of doing and acting, as living people 'imitate' other living people by performing fictional actions that are nonetheless realistic and probable. It is worth recalling that the term 'invention' as used by Shakespeare and other Renaissance

> The human mortals want their winter cheer.
> No night is now with hymn or carol blessed.
> Therefore the moon, the governess of floods,
> Pale in her anger, washes all the air,
> That rheumatic diseases do abound;
> And through this distemperature we see
> The seasons alter: hoary-headed frosts
> Fall in the fresh lap of the crimson rose,
> And on old Hiems' thin and icy crown
> An odorous chaplet of sweet summer buds
> Is, as in mock'ry, set. The spring, the summer,
> The childing autumn, angry winter change
> Their wonted liveries, and the mazèd world
> By their increase now knows not which is which;
> And this same progeny of evils comes
> From our debate, from our dissension –
> We are their parents and original.
>
> (2.2.101–17)

Again Shakespeare bookends the passage with a clear causal agent, assigning it to the quarrel between Oberon and Titania, but this glimpse of order should not distract us from the fact that the rest of the passage pulls centrifugally away from any notion of cause and the underlying notions of subjectivity and unity – of active unities acting on other unities in a series – that 'cause' always implies. Nor should the 'distemperature' of nature be understood as a model of 'structural' cause, which simply displaces the notion of subject onto structure, but rather as infection: the 'contagious fogs' at the opening of the speech are neither subjects nor unities of any kind but clouds of an un-

writers to describe the poetic process meant 'to find' or to 'discover' (from the Latin *invenire*); the poet's imaginative faculty, combined with his memory, allowed him to 'invent' for his poems and plays new 'devices,' as they were often called, another term that had both intellectual and mechanical meanings in the sixteenth century. This imaginative and poietic creativity was the measure of a poet's ingenuity, a word derived from *ingenium*, the same Latin root as 'genius' and 'engineer': to write poetry was to work like a technician in its original etymological sense, using 'art' and linguistic tools such as metaphor to create entities that simulated nature and even went beyond conventional notions of what nature could naturally produce.

Can we not make the very same claims for modern science? Scientific inquiry in general, especially the process of experiment and 'discovery,' depends on modes of representation that can legitimately be called mimetic and that are important to articulate if we wish to understand the claims to moral authority that are often made in the name of scientific knowledge. After all, in the words of the sociologist of science Karin Knorr-Cetina, 'sign-creating technologies, in so far as they turn out verbal renderings, visual images, or algorithmic representations of objects and events, seem to be present in all sciences' (41). Socrates himself would be especially intrigued by high energy physics, where 'the construction of objects as "signatures" and "footprints" of events, rather than as the events themselves, shapes the whole technology of experimentation' (41). Here, as Knorr-Cetina argues, experiments process the signs that laboratories make (42); high energy physics 'moves in the shadowland of . . . negative images of the world – in a world of signs and often fictional reflections,

designated fatal force that transmits itself without clear agent or trajectory. The moon appears briefly as a subject-image, but this effect is muted by the equally diffusive images of 'washing the air' and of 'rheumatic diseases,' which strain against any clear notion of causation (like the more positive 'odorous' later in the passage). The seasons suddenly 'alter' in a middle mode somewhere between active and passive tense, their tokens juxtaposed with one another through a minimal passive ('is set'); if the 'chiding autumn' and 'angry winter' burst forward as personified agents toward the end of the passage, they do so only to sink again into indistinction two lines later – it is not merely the idea of 'season' that is undone but the idea of individuation itself that threatens to dissolve into a tempest of force and constantly mutating elements.

Magic and *mimesis*

If myth offered Shakespeare one resource for examining the mysterious and inhuman operations of nature, the magic and astrology of his period provided another. For writers such as John Dee, Henry Cornelius Agrippa, Giovanni Battista Della Porta, and Marsilio Ficino, the 'occult' properties of magic and astrology included all the natural processes that remained hidden from the naked eye: the metaphysical forces of the stars and planets, the immanent powers of magnets or medicinal compounds, the mechanical power of devices and machines. 'So there are in things, besides the elementary qualities which we know,' Agrippa writes, 'other certain inbred virtues created by nature, which we admire, and are amazed at, being such as we know not, and indeed seldom or never have seen' (32). For this

of echoes, footprints, and the shimmering appearances of bygone events' (46):

> These objects are in a very precise sense 'unreal' – or, as one physicist described them, 'phantasmatic' (*irreale Gegenstände*); they are too small ever to be seen except indirectly through detectors, too fast to be captured and contained in a laboratory space, and too dangerous as particle beams to be handled directly. Furthermore, the interesting particles usually come in combination with other components that mask their presence. Finally, most subatomic particles are very short-lived, transient creatures that exist only for a billionth of a second. Subject to frequent metamorphosis and to decay, they 'exist' in a way that is always already past, already history. (48)

These 'inscriptions' of the laboratory 'translate' the world into an enduring form, to use the terminology of Latour, and without these different 'translations' there would be no data, no fact, no experiment, and no theory. Whereas once the role of the *onomastikos* was played by the poet, giving 'a local habitation and a name' to 'the form of things unknown,' as Theseus objected in *A Midsummer Night's Dream*, now the *onomastikos* has become the scientist who produces new things that we have never seen before, naming them and integrating them into preexisting systems of explanation and evidence. Facts mean nothing until they have been digested into a system: they are not even 'facts' until this incorporation has occurred and relevant 'information' has been sorted from irrelevant 'noise.' As much as a play or a poem, an experiment depends on a shared, communal

reason all magic is perfectly natural in Agrippa's eyes: 'for every day some natural thing is drawn by art, and some divine thing drawn by nature, which the Egyptians seeing, called Nature a magicianess, i.e. the very magical power itself' (110). Many aspects of the 'occult sciences,' as Agrippa called them, can seem surprisingly modern and include phenomena we would describe as agricultural, chemical, or industrial: the grafting of plants and the incubation of seeds and eggs, the rising of bread, oils that burn on water, the smelting of metals, even the process of digestion. Historians of science such as William Eamon, Pamela Smith, and William Newman have shown how much the fields of magic, astrology, and alchemy contributed to the emergence of modern experimental method, since they provided a language and a set of practical methods for inquiring into the nature of physical substance and its behavior under artificially induced conditions, including problems of composition and mixture, of degeneration and decay, of instrumentality, causation, and force.

At the same time, however, the works of Agrippa and Ficino clearly demonstrate how far from our modern notions of 'science' occult philosophies could be, and this is because their methods are as ritualistic and symbolic as they are mechanical or physical. Indeed, we may go a step further and describe all of the occult sciences as *mimetic*, since they presume an underlying principle of resemblance and grant to imitation a causal power. One of the most important aspects of early modern philosophies of life, at least as these are articulated within the occult sciences, lies in the fact that their vital principle is mimetic and imitative rather than chemical or molecular. Drawing an image allows the magus to draw down the virtues of the thing imitated

knowledge of conventions and codes in order for it to be meaningful; indeed, it is the interpretation of the experiment, rather than the data it generates, that is most significant in any laboratory.

Nor can we maintain that the field of molecular biology is any less subject to signs, language, and *mimesis* than the field of high energy physics (as Knorr-Cetina herself suggests). For the problem of onomapoetics in science is particularly well-illustrated by the history of the gene itself. 'The power of words,' Evelyn Fox Keller has observed:

> derives from a relation to things that is always, and of necessity, mediated by language-speaking actors. Like the rest of us, scientists are language-speaking actors. The words they use play a crucial (and, more often than not, indispensable) role in motivating them to act, in directing their attention, in framing their questions, and in guiding their experimental efforts. By their words, their very landscapes of possibility are shaped. (*Century*, 139)

For a long time, Keller points out, 'genes were hypothetical entities' (*Century*, 19) rather than 'real' or 'factual' things; their structure was in many ways projected by analogy from the meaningful epistemic units of other scientific fields: 'just as atoms and molecules provided the fundamental units of explanation in physics and chemistry, so too would particulate hereditary elements serve as the fundamental units of biological explanation' (*Century*, 18). Charles Darwin provides a famous example: preoccupied with defining the unit of inheritance, Darwin turned to the ancient theory of 'pangenesis' in order to solve some of the enduring logical problems in the theory of evolution. Some

and to communicate that virtue to the beholder via the eye; to speak the name of a planet is to invoke its power; to wear an amulet or precious stone is to attract the sympathetic rays of the stars that are thought to govern the substance around your neck, all because of an essential affinity between the name, the image, the substance, and the referent. For Agrippa, the life-giving 'virtue' of the world-spirit spreads through a process of self-replication 'by a certain likeness, and aptness that is in things amongst themselves toward their superiors' (106); all magic consists in 'the attracting of like by like, and of suitable things by suitable' (110). A similar notion of imitation is precisely what allows Ficino to distinguish between a lawful and therapeutic use of astrological images and an unlawful and heretical one: 'be warned beforehand not to think we are speaking here of worshipping the stars, but rather of imitating their power and thereby trying to capture them through imitation' (356–7; translation modified). Like Agrippa, Ficino everywhere stresses the beneficent and natural basis of his magic and is manifestly nervous about how his notion of 'daemons' – life-giving forces that naturally occur in the universe and that function through imitation – might be misinterpreted.

We may suspect that the mimetic aspect of the occult sciences captured Shakespeare's imagination as he sought analogies for the mysterious process of poetic creation. In *A Midsummer Night's Dream*, too, 'nature' includes the actions of daemons who can blend and mix substances at will, sometimes with a therapeutic purpose but often for malicious ones. The confrontation between Titania and Oberon seems to involve control over a mysterious life-giving power that is very like the *spiritus mundi* Ficino describes, a 'body living in every part, as is

mechanism must account for the transmission of both hereditary and acquired characteristics and also allow for variation within species, Darwin reasoned; the theory of pangenesis suggested that hereditary characteristics were contributed by all of the cells in the body to the gametes (the reproductive cells), which in turn passed them from generation to generation. 'Gemmules' was Darwin's term for this critical unit of inheritance, and pangenesis the theory that might explain their dissemination.

But Darwin recognized that his notion remained all-too hypothetical and, with *A Midsummer Night's Dream* in mind, he wrote to his friend and protegé the biologist George John Romanes, expressing his hope that he would 'some day . . . convert an "airy nothing" into a substantial theory.' His offhand invocation of Theseus should remind us that scientific knowledge owes more to the operations of poetry and theater than most scientists are willing to recognize. As the philosopher of science Paul Feyerabend has written, 'Knowledge . . . is not a series of self-consistent theories that converges towards an ideal view; it is not a gradual approach to the truth. It is rather an ever increasing *ocean of mutually incompatible alternatives*, each single theory, each fairy-tale, each myth that is part of the collection forcing the others into greater articulation and all of them contributing, via this process of competition, to the development of our consciousness' (21).

Mimesis and science

The more we consider scientific experimentation, indeed, the more it appears to operate like a grand *coup de théâtre*. It is intentional: it has been shaped and organized to communicate

evident from motion and generation.' 'The philosophers of India deduce its life from the fact that it everywhere generates living things out of itself' (255), Ficino writes, and 'the life of the world, innate in everything, is clearly propagated into plants and trees, like the body-hair and tresses of its body . . . the world is pregnant with stones and metals, like its bones and teeth' (289). Titania is also worshipped in India:

> The fairyland buys not the child of me.
> His mother was a vot'ress of my order,
> And in the spicèd Indian air by night
> Full often hath she gossiped by my side,
> And sat with me on Neptune's yellow sands,
> Marking th'embarkèd traders on the flood,
> When we have laughed to see the sails conceive
> And grow big-bellied with the wanton wind,
> Which she with pretty and with swimming gait
> Following, her womb then rich with my young squire,
> Would imitate, and sail upon the land
> To fetch me trifles, and return again
> As from a voyage, rich with merchandise.
> But she, being mortal, of that boy did die;
> And for her sake do I rear up her body;
> And for her sake I will not part with him.
>
> (2.1.122–37)

In Titania's account, the pregnancy of her votive-priestess has no clear cause and no paternal agent but seems instead to result from the mimetic principles of acting, dance, and metaphor, which Shakespeare imagines as a cosmic living force. Through

an idea about 'things that could be the case,' as Aristotle had defined *mimesis* in his *Poetics*. For this reason, the experiment is hypothetically driven, like the fiction or *mythos* of the playwright; like Aristotle's playwright, the scientist works, at best, in generalities. No less a figure than Francis Crick draws a fundamental distinction between physics and biology on this very point:

> The basic laws of physics can usually be expressed in exact mathematical form, and they are probably the same throughout the universe. The 'laws' of biology, by contrast, are often only broad generalizations, since they describe rather elaborate chemical mechanisms that natural selection has evolved over billions of years. Biological replication, so central to the process of natural selection, produces many exact copies of an almost infinite variety of intricate chemical molecules. There is nothing like this in physics or its related disciplines. That is one reason why, to some people, biological organisms appear infinitely improbable. (*What Mad Pursuit*, 5)

Both Knorr-Cetina and Robert Crease have argued that experimental procedures are best understood as kinds of theatrical performance. Models that simulate actual circumstances at reduced scale, computer simulations, psychological and economic experiments about group interaction or decision making: 'in most respects,' writes Knorr-Cetina, 'the laboratory is a virtual space [that is] coextensive with the experiment. Like a stage on which plays are performed from time to time, the laboratory is a storage room for the stage props that are needed when social life is instantiated through experiments' (35). For this reason we should never describe a scientific experiment as

a metaphor suggested by a formal resemblance, the sails of ships suddenly balloon into a belly that has been inseminated by the wind; the priestess is first a spectator to this show of metaphoric transfer, then an actor who joins the show by embodying those principles, 'sailing' upon the land in a natural dance that transforms noun into verb while uttering a bellowing laugh.

Titania's description calls our attention to the fact that theater is always an embodied medium, one in which the body becomes the source of all subsequent formal thought, as Elizabeth Sewell has argued:

> If, for the living individual, the body is the original generator of forms, first its own form in structure and behavior, then forms which are in varying degrees separated from itself and which accordingly offer the mind-body scope for *its* formalizing tendencies, it may be true to say that all formal activity in the human mind has its origin and roots always in the physical. The mind-body may generate forms as languages or terms for metaphoric activity by which to understand itself and its experience; but all form, no matter how apparently abstract and intellectual, may never lose its connection with, its message for, the body. It seems possible that all forms observed by or constructed by the so-called mind are *Gestalten* or figures or forms in the recurring double sense of all those terms: that a figure is always an image; that a form, how abstract soever, calls forth from the body a physical response, is perceived as an image, if that is the right word, by the body which is the source of all forming activity. (36–7)

'natural': it is an artificially composed arrangement of events that often would never occur in so-called nature and whose 'naturalness' can only be inferred backwards and retrospectively from the entirely unnatural circumstances in which they appear. Bacon recognized this aspect of scientific method clearly: 'Proteus never changed shapes till he was straightened and held fast,' he observes, 'so the passages and variations of nature cannot appear so fully in the liberty of nature as in the trials and vexations of art' (*Advancement*, 77). Already at the dawn of 'modern' science, Bacon had grasped that the new *organon* he was forging consisted in nothing less than identifying operations in Nature that had never been seen before and, like a new species of poet, giving them new names: 'attraction, repulsion, attenuation, conspissation, dilatation, astriction, dissipation, maturation, and the like' (*New Organon*, 63).

Experimental procedure is thus doubly 'dramatological': in the theatrical sense because it depends on physical movement, gesture, on carefully staged circumstances, on managed relationships with bodies, props, and models; and in the strong philosophical sense because the experiment constitutes performatively the events that it purports to reveal. Many scientists dislike these arguments, because they prefer to think of natural structures as waiting passively and invisibly for discovery by instruments that reproduce the phenomena without mediation or loss. As Latour has argued, however, the things that we 'know' in a scientific sense are thinkable to us only because of the models, images, diagrams, formulas, rules, and equations that are inseparable from them, and this imbrication of inscriptions and things is no less true of common-sense perception than of highly technical conditions. Feyerabend puts the problem plainly:

For Sewell, the 'body' is more than simply flesh or substance but should rather be understood as a hybrid body-mind or mind-body endowed with a mode of apprehension that we reduce if we describe it merely as 'sensation' or as 'experience,' to use the terms of a philosophical tradition – call it 'Platonic' or 'Cartesian' – that would separate the body from the mind and then forge its epistemological and ontological categories on this distinction. For Shakespeare, theater provided a model for the type of body-mind that Sewell describes, and with it an entire approach to thought, feeling, and imaginative creativity that moves in an entirely different direction from the Platonic and Cartesian philosophical traditions: we find not a mute, passive body subordinated to abstract reflection but rather a generative and recursive body-mind out of which all knowledge and all expression emerges. And since this body-mind, as a temporal mode of being, is always mutating and changing, it implies a necessary mortality, much the way the 'mortal' nature of the priestess is sublimated into the new life of a poetic figure upon the birth of her child, which is also the cause of her own death.

On the one hand, therefore, Titania's speech emphasizes a natural foundation in the body for all language, all form, and all poetic image, a spontaneous creative process that becomes especially visible in the theater, composed as it is out of many nested elements: a turn of phrase releases a word into the air to be captured by a flick of the hand and spinning finger, while a foot slaps and the body pivots with a comically distended belly to 'sail' across the stage. But at the same time the speech calls attention to the fact that the 'body' generating these poetic figures is itself already a kind of figure, an image narrated on

[When evaluating any scientific claim] . . . we must become clear about the nature of the total phenomenon: appearance plus statement. There are not two acts – one, noticing a phenomenon; the other, expressing it with the help of the appropriate statement – but only one, viz. saying in a certain observational situation, 'the moon is following me,' or, 'the stone is falling straight down' [two famous examples considered by Galileo]. We may, of course, abstractly subdivide this process into parts . . . But under normal circumstances such a division does not occur; describing a familiar situation is, for the speaker, an event in which statement and phenomenon are firmly glued together.

This unity is the result of a process of learning that starts in one's childhood. From our very early days we learn to react to situations with the appropriate responses, linguistic or otherwise. The teaching procedures both shape the 'appearance,' or 'phenomenon,' and establish a firm connection with words, so that finally the phenomena seem to speak for themselves without outside help or extraneous knowledge. They are what the associated statements assert them to be. (57)

In a similar vein, the anthropologist Michael Taussig points out that even perception itself, the basis of all empirical method, is fundamentally mimetic: the stimulus of the sun's rays on the organ of the eye produces a mental image of the sun, a repetition or copy that is also, at the same time, an original sensation. How could we 'see' the force and physics of sunlight without these types of representations? A graph, a spectrometer, a rainbow – all are representations of this thing we call 'sunlight,' which is itself, of course, a word that imitates other words and composes

stage by a male actor, who, in speaking, assembles the character of Titania by evoking the qualities associated with her. At this level of the speech, we find that the so-called 'body' is no more coherent and unified, no more essential or natural than the poetic forms and figures that it would putatively generate. In this way Titania's speech offers one of the best examples in the play of the way in which *A Midsummer Night's Dream* dissects the theater's ability to create life through artificial means, from the smallest unit to the largest: vocable, syllable, word, metaphor, image, line, sentence, phrase, speech, character, prop, group, 'scene,' 'plot,' to play itself – the component cells, as we could call them, of a self-replicating organism that lived through artifice and grew by assemblage.

All their minds transfigured so together

Reading through Agrippa and Ficino, one is struck by how many resonances there are between the occult vision of nature and the world of *A Midsummer Night's Dream*, not least because of the attitude toward language shared by poets and magicians, both of whom granted to words and names an active, causal power. Orpheus, after Apollo the most significant mythical source of poetry and music, was also one of the most important figures in early modern myths of the ancient magi, second, in some accounts, only to Hermes himself. 'They that desire further examples' of the power of words and names, suggests Agrippa, should 'search in into the hymns of Orpheus, than which nothing is more efficacious in natural Magic' (156). 'In natural magic nothing is more efficacious than the Hymns of Orpheus,' affirmed Giovanni Francesco Pico Della Mirandola,

itself out of them. 'To get hold of something by means of its likeness. Here is what is crucial in the resurgence of the mimetic faculty,' Taussig argues (21): 'can't we say that *to give an example, to instantiate, to be concrete*, are all examples of the magic of mimesis wherein the replication, the copy, acquires the power of the represented?' (16). The mystery of experiment lies in this reiterative, mimetic quality that constitutes any moment of 'discovery': a phenomenon that is utterly new gradually assembles itself out of a constellation of variables that we already recognize and that it seems somehow to resemble.

An excellent example of how *mimesis* makes possible the passage from airy nothing to concrete something can be found in Watson and Crick's accounts of the 'discovery' of the double helix. Both men wrote famous memoirs, and both emphasize the collaborative, hypothetical, and accidental nature of scientific research. 'It took over twenty-five years for our model of DNA to go from being only rather plausible, to being *very* plausible . . . and from there to being virtually certainly correct,' Crick wrote. 'Even then it was correct only in outline, not in precise detail . . . The establishment of the double helix could serve as a useful case history, showing one example of the complicated way theories become "fact"' (*What Mad Pursuit*, 73–4). Eager to beat Linus Pauling to the discovery of the structure of DNA, and distracted, Lysander-like, by thoughts of pretty foreign girls, Watson 'started wondering whether each DNA molecule consisted of two chains with identical base sequences held together by hydrogen bonds between pairs of identical bases' (*Double Helix*, 184). It was only after his colleague Jerry Donohue pointed out that 'I had chosen the wrong tautomeric forms of guanine and thymine' and that 'for years organic chemistry textbooks were littered with pictures

in a passage that might well have been written as a gloss on Oberon's 'experiment' with the potion in *A Midsummer Night's Dream*:

> the names of the gods of which Orpheus sings are not those of deceiving demons, from whom comes evil and not good, but are names of natural and divine virtues distributed throughout the world by the true God for the advantage of man, if he knows how to use them. (Yates, *Bruno*, 89)

Thus when Sidney compares the power of Orpheus to move trees with Stella's ability to 'charm' men's ears in his lyric 'If Orpheus voice had force to breath such music's love'; when, in his *Defence*, he calls Orpheus 'Father in learning' for all historians; when he argues that the poet ranges like the astrologer 'only within the zodiac of his own wit' to make monstrous, unnatural creatures; or when he claims that the poet, like the alchemist, discerns the secret processes of natural life and harnesses them to fabricate a series of 'golden' forms, he places poesy directly within an occult tradition that Shakespeare exploits to the fullest in the woods outside of Athens.

The most powerful tool wielded by the magician, after all, was the word, especially when words were combined into sentences or sung in verse form, as the poets do. 'Words therefore are the fittest medium betwixt the speaker and the hearer,' says Agrippa, 'carrying with them not only the conception of the mind but also the virtue of the speaker with a certain efficacy to the hearers, and this oftentimes with so great a power, that oftentimes they change not only the hearers but also other bodies and things that have no life' (211). 'There is no less virtue in words and the names of

of highly improbably tautomeric forms over their alternatives on only the flimsiest of grounds' (189–90) that Watson was able to propose a hypothetical model that the larger scientific community (at first the very small community of the Cavendish and King's College laboratories) could agree upon.

The famous story was in many ways a struggle for authority between two different methods of *mimesis*: the three-dimensional models of wood and metal built by the Cavendish Laboratory at Cambridge, on the one hand, and the x-ray crystallography photographs taken by Rosalind Franklin at Maurice Wilkins's lab at King's College, London, on the other. The model allowed Watson and Crick (and in time the entire scientific community) to visualize a chemical structure that no one had yet grasped, and for this reason their models were helixes of pure poetry, in the truest sense of the word: they had made a new thing. Until the model was built, the structure of the DNA strand was a hypothetical entity which might have several different forms, and it would be perfectly fair to say that the double helix simply didn't exist in the same way that it did after the model was built. And both Crick and Watson were keenly aware that the models could be as misleading as they could lead to a valuable insight. But because they were willing to build a type of model that Wilkins and Franklin dismissed, they were able to solve a problem that many of their colleagues regarded as the key to the definition of biological life.

Although the wild promise of recombinant DNA technology remained unanticipated, finally the mysterious 'transforming factor,' as Oswald Avery, Colin MacLeod, and Maclyn McCarty had called it in one of the earliest experimental isolations of DNA at the Rockefeller Institute, showed a regular structure. This

things,' he continues, 'but greatest of all in speeches and motions (151) . . . They say that the power of enchantments and verses is so great, that it is believed they are able to subvert almost all nature, as saith Apuleius, that with a Magical whispering, swift Rivers are turned back, the slow Sea is bound, the Winds are breathed out with one accord, the Sun is stopped, the Moon is clarified, the Stars pulled out, the day is kept back, the night is prolonged' (157). This is why Puck, Titania, and Oberon speak to one another in verse as they dance together at the end of the play in order to reset the dislocation of the seasons (4.1.92–101); why the faeries sing and dance to protect against the threats of both nature and art ('Weaving spiders'; 'Beetles black'; 'Worm nor snail'; 'Nor spell nor charm' [2.2.20–30]); and why Oberon and Puck must chant their spells while applying the potion (to Titania [2.2.33–40]; to Lysander [2.2.72–89]; to Demetrius [3.2.102–9]) or when removing it again (from Lysander [3.2.448–64, with variation]; from Titania [4.1.70–4]). Ficino insists on the same point:

> They hold that certain words pronounced with a quite strong emotion have great force to aim the effect of images precisely where the emotions and words are directed. And so, in order to bring two people together in passionate love, they used to fashion an image when the Moon was above the horizon and was coming together with Venus in Pisces or Taurus, and they followed many precise directions involving stars and words which I will not tell you, for we are not teaching philters but medicine. (355)

structure provided a clear indication of how genetic material passed from cell to cell and across generations: it was *copied* through a process whereby two strands of simple base pairs, running parallel in a reverse double helix, unwound and began generating their complementary molecular units. 'It has not escaped our notice,' Watson and Crick wrote in their first publication of their findings in the journal *Nature*, 'that the specific pairing we have postulated immediately suggests a possible copying mechanism for the genetic material' (*What Mad Pursuit*, 66). If the sentence was 'coy,' Crick later explained, this was because it represented a 'compromise, reflecting a difference of opinion. I was keen that the paper should discuss the genetic implications' of the discovery, but 'Jim was against it. He suffered from periodic fears that the structure might be wrong and that he had made an ass of himself' (*What Mad Pursuit*, 66).

An ass, indeed – perhaps such mutations are an inevitable risk when one works with a new 'transforming substance.' Crick, however, was delighted, and he decided to commemorate the event by renaming his house 'The Golden Helix,' turning for his inspiration to the very story about man-transformed-into-ass that had prompted Shakespeare to write *A Midsummer Night's Dream*:

> For a number of years after that, things were fairly quiet. I named my family's Cambridge house in Portugal Place 'The Golden Helix' and eventually erected a simple brass helix on the front of it, though it was a single helix rather than a double one. It was supposed to symbolize not DNA but the basic idea of a helix. I called it golden in the same way that Apuleius called his story 'The Golden Ass,' meaning beautiful. (*What Mad Pursuit*, 78)

For Ficino, charms that are sung are especially effective because they distill the essence of universal life: 'the very matter of song, indeed, is . . . air, hot or warm, still breathing and somehow living; like an animal, it is composed of certain parts and limbs of its own and not only possesses motion and displays passion but even carries meaning like a mind, so that it can be said to be a kind of airy and rational animal' (359).

This living quality that is quasi-demonic in its power forms one of the most important links between the occult sciences and theater: 'remember that song is a most powerful imitator of all things,' Ficino writes, in words echoed by the anti-theatricalists of Shakespeare's period, 'it imitates the intentions and passions of the soul as well as words; it represents also people's physical gestures, motions, and actions as well as their characters and imitates all these and acts them out so forcibly that it immediately provokes both the singer and the audience to imitate and act out the same things' (359). This is also why Egeus can complain at the very opening of the play that Lysander has:

> bewitched the bosom of my child.
> Thou, thou, Lysander, thou hast given her rhymes,
> And interchanged love tokens with my child.
> Thou hast by moonlight at her window sung
> With feigning voice verses of feigning love,
> And stol'n the impression of her fantasy
> With bracelets of thy hair, rings, gauds, conceits,
> Knacks, trifles, nosegays, sweetmeats – messengers
> Of strong prevailment in unhardened youth.
>
> (1.1.27–35)

A remarkable reference point for a Nobel-Prize winning biologist whose first prize, won at school when he was a boy, was for collecting wild flowers, Oberon-like, in the countryside outside his Midlands home (*What Mad Pursuit*, 9).

As Crick maintained throughout his life, scientific discovery required not just knowledge or discipline or method but constant, collaborative conversation and that utterly unpredictable spark of creative imagination that Plato and Theseus had associated with the poet. And he emphasized the point by borrowing the title of his memoir, *What Mad Pursuit: A Personal View of Scientific Discovery*, from one of the most famous poems in English, Keat's 'Ode on a Grecian Urn':

> Thou still unravished bride of quietness,
>   Thou foster-child of silence and slow time,
> Sylvan historian, who canst thus express
>   A flowery tale more sweetly than our rhyme:
> What leaf-fringed legend haunts about thy shape
>   Of deities or mortals, or of both,
>   In Tempe or the dales of Arcady?
> What men or gods are these? What maidens loath?
>   What mad pursuit? What struggle to escape?
>   What pipes and timbrels? What wild ecstasy?
>   (1–10)

Keats's lines might just as well describe Shakespeare's forest outside of Athens as they do the 'Golden Helix,' famous in the 1960s for the parties thrown by Crick and his wife Odile at which guests were invited to sketch nude models and watch pornographic films screened artfully in reverse.

Egeus accuses Lysander of acting like a poet as well as like a magician, distorting Hermia's judgment by observing all the rituals necessary to love-magic.

And what if Oberon were cast as a scientist made mad by myth and song? While on a research trip in the field to observe dolphins and listen to the 'dulcet and harmonious breath' of mermaids (2.1.151), Oberon has found a 'herb' (2.1.169) that grows naturally in the wild but which has undergone an accidental mutation through contact with the mythic force of Cupid's arrow:

> Yet marked I where the bolt of Cupid fell.
> It fell upon a little western flower –
> Before, milk-white; now, purple with love's wound;
> And maidens call it 'love-in-idleness'.
> Fetch me that flower; the herb I showed thee once.
> The juice of it on sleeping eyelids laid
> Will make man or woman madly dote
> Upon the next live creature that it sees.
>
> (2.1.148–72)

Nourished by the blind arbitrariness of a desire that has the power to transform anything it touches, the flower is the trace not only of Cupid's bow but of the literary sources that Shakespeare transforms as he incorporates them: the Pyramus and Thisbe myth (in which the fruit of the mulberry tree is blackened by Pyramus's blood), crossed with the story of Cupid and Psyche, as recounted to a young maiden imprisoned at the center of Apuleius's *Golden Ass* and reported by the narrator Lucius, who has himself been transformed into a donkey with

Love's stories written in love's richest book

Crick seems to have recognized that he had not only 'discovered' the structure of DNA: he had produced an extremely powerful poetic symbol of life in general that would henceforth govern all future inquiry into genetic reproduction and that would soon lead him to his next major contribution to modern molecular biology: the cracking of the genetic code. Crick's subsequent work returns us to Dr George Johnson's analogy in the *St Louis-Post Dispatch* with which we began: genetic engineering is like 'inserting a phrase from Shakespeare into a line of code in a computer program.' Some readers might find a comparison with the thesaurus more appropriate (genetic engineering is like deliberately substituting one word for another in the interest of improving style); others might think of a compositor's error in a printed book (and in fact printed books are a commonplace metaphor in discussions of DNA, as we shall see). Nevertheless, Johnson's analogy introduces us to a second reason why the field of biotechnology should be understood as a new permutation in the long history of the concept of *mimesis*. For the notion of 'code' joins a cluster of concepts, including that of the 'program,' of alphabets, of typeset printing, and of 'translation,' whereby the DNA generates the RNA that carries the information outside the nucleus to generate a new inverse string of base-pairs, all of which imagine genetic life as a process of copying, self-duplication, and as so many particular forms of writing. The notion of life as 'code' thus recombines, we might say, with the other mimetic strand I have identified as characteristic of scientific inquiry in general, namely Latour's notion of 'inscription' and of the many types of graphic 'translation' necessary to produce scientific knowledge.

human thoughts and appetites. Like any mythic substance (or any experimental compound) the potion's power is unpredictable: 'This flower's force in stirring love' (2.2.75) also generates the 'hateful imperfection' (4.1.62) that clouds Titania's eyes; if it produces passionate devotion, it also results in her debasement and humiliation at the hands of Oberon, who 'taunts' her 'at my pleasure' while 'she in mild terms begged my patience' (4.1.56–7).

No sense is more susceptible to magic than the sense of sight, Agrippa argues, 'because it perceives more purely and clearly than the other senses' (134). 'There is therefore a wonderful virtue, and operation in every herb' (39), he explains, which 'can multiply, transmute, transfigure, and transform' (134) by penetrating to the imagination. Under the effects of the potion, Lysander and Demetrius gaze on Helena with new eyes, endowed with a sudden, sonnet-like eloquence; they charge through the forest in jealous rage, chasing Puck's calls and ventriloquizing effects. 'This is the cause why maniacal, and melancholy men believe they see, and hear those things without, which their imagination doth only fancy within,' Agrippa writes, 'hence they fear things not to be feared, and fall into wonderful and most false suspicions, and fly when none pursueth them, are angry, and contend, no body being present, and fear where no fear is' (134–5). With an irony that anyone but a lover could hear, Lysander believes himself to be utterly possessed of his senses and to be compelled by logic rather than by the potion or by his desire:

> Not Hermia but Helena I love.
> Who will not change a raven for a dove?

As both Lily Kay and Evelyn Fox Keller have shown, the concepts of 'code,' 'program,' 'information,' and 'translation' had become dominant metaphors for the operation of DNA since at least the mid–1940s, when the physicist Erwin Schrödinger referred to DNA as the 'program' of the cell in his book *What is Life?*, a work that was influential on both Watson and Crick and on many others working in the nascent field of molecular biology. The geneticist and Nobel laureate George Beadle went so far as to compare the decoding of DNA to the discovery of the Rosetta Stone:

> What has happened in genetics during the past decade has been the discovery of a Rosetta stone. The unknown language was the molecular one of DNA. Science can now translate at least a few messages written in DNAese into the chemical language of blood and bone and nerves and muscle. One might also say that the deciphering of the DNA code has revealed our possession of a language much older than hieroglyphics, a language as old as life itself, a language that is the most living language of all – even if its letters are invisible and its words are buried deep in the cells of our bodies. (Kay, 17)

Beadle joined a long tradition of scientists who regarded their method as one of reading and interpreting the so-called 'book of nature,' a tradition that received one of its most famous formulations from Galileo in 1623, only seven years after Shakespeare's death, as we have seen above, when he spoke of mathematics as the alphabet of the universe.

The first step toward the cracking of the genetic code was to identify its meaningful units, and Crick, too, turned to the

> The will of man is by his reason swayed,
> And reason says you are the worthier maid.
> Things growing are not ripe until their season,
> So I, being young, till now ripe not to reason.
> And, touching now the point of human skill,
> Reason becomes the marshall to my will,
> And leads me to your eyes, where I o'erlook
> Love's stories written in love's richest book.
> (2.2.119–28)

Helena appears more beautiful; she is therefore the more worthy object; it is the duty of a mature young man to woo accordingly. And Lysander finds confirmation for his argument in nature's own peculiar 'art,' written 'transparently' across Helena's body: the blushing language of the skin, perhaps, or the morse code of a fluttering pulse, or eyes like a microchip, encoding an entire bibliography of desire that only the frantic lover (or the frenzied poet) can access.

The problem of 'judgment' is a fundamental one in *A Midsummer Night's Dream*, and Shakespeare seeds the term throughout the play in both its early modern senses of 'knowing through sense perception and the discrimination of difference' as well as 'evaluation, criticism, and rule of law.' The two meanings neatly combine in Theseus's speech to Hermia early in the play:

THESEUS: Demetrius is a worthy gentleman.
HERMIA: So is Lysander.
THESEUS: In himself he is;
    But in this kind, wanting your father's voice,
    The other must be held worthier.

metaphor of mechanical reproduction and of print in order to explain the difficulty facing those who would read the as-yet-undeciphered sequence. 'The amino acids . . . are just like the letters in a font of type,' Crick explained:

> The base of each kind of letter from the font is always the same, so that it can fit into the grooves that hold the assembled type, but the top of each letter is different, so that a particular letter will be printed from it. Each protein has a characteristic number of amino acids, usually several hundred of them, so any particular protein could be thought of crudely as a paragraph written in a special language having about twenty (chemical) letters . . . And now we can approach the baffling problem that appeared to face us. If genes are made of protein, it seemed likely that each gene had to have a special three-dimensional, somewhat compact structure. Now, a vital property of a gene was that it could be copied exactly for generation after generation, with only occasional mistakes. What we were trying to guess was the general nature of this copying mechanism. Surely the way to copy something was to make a complementary structure – a mold – and then to make a further complementary structure of the mold, to produce in this way an exact copy of the original. This, after all, is how, broadly speaking, sculpture is copied. But then the dilemma arose: It is easy to copy the outside of a three-dimensional structure in this way, but how on earth could one copy the inside? The whole process seemed so utterly mysterious that one hardly knew how to begin thinking about it. (*What Mad Pursuit*, 34–5)

HERMIA: I would my father looked but with my eyes.
THESEUS: Rather your eyes must with his judgement look.
(1.1.52–7)

To submit to the 'judgment' of paternal authority is not simply to extinguish desire but to disregard one's own senses, or to allow authority to distort perception, like a kind of magic charm. The speech establishes an implicit analogy between the legitimate (or legit*imized*) magic of law and justice, on the one hand, and the illegitimate magic of spells and potions, on the other, a comparison reinforced by the theatrical doubling of Theseus and Oberon. The same analogy returns again at the end of the play in a detail that often goes unremarked by audiences: the potion is never in fact lifted from Demetrius's eyes, implying that a life lived under the law, and especially erotic life under the law, is in fact a life of deception, illusion, distortion, and submission to an arbitrary norm.

As Puck's experiment with Lysander demonstrates, the power of the potion is to induce perversion in its etymological sense of a 'turning away' from a 'proper' erotic object toward an 'improper' one. Certainly the implication is that the potion will provoke Titania 'to do those things which men that are awake cannot, or dare not do' as Agrippa puts it, the erotic congress between maiden and ass that Lucius in Apuleius's *Golden Ass* recounts with relish. And just as it dawns on us that *A Midsummer Night's Dream* is really a comedy about deviations in sexual object and sexual aim, as Freud might have described it, Bottom and company enter to remind us that we are watching a theatrical performance where the parts of Hermia and Helena (as well as Thisbe) are all being played by men, who direct all the watching and wooing and

The key to the problem lay in the process of natural selection, which Crick never tired of defending vigorously against religious ideology:

> What, then, are the basic requirements for natural selection to work? We obviously need something that can carry 'information' – that is, the instructions. The most important requirement is that we should have a process for exact replication of this information. It is almost certain that, in any process, some mistakes will be made, but they should occur only rarely, especially if the entity to be replicated carries a lot of information . . . The second requirement is that replication should produce entities that can themselves be copied by the replication process or processes. Replication should not merely be like that of a printing press, when master plates make many copies of a newspaper but each newspaper cannot, by itself, produce further copies of either the press or the newspaper . . . The third requirement is that mistakes – mutations – should themselves be capable of being copied, so that useful variation can be preserved by natural selection. (*What Mad Pursuit*, 27)

With the advent of a notion of life as code and information, the principles of logic and of mathematical combination began to replace any lingering vestiges of vitalism or of a 'design' thesis, which Crick himself enthusiastically attacked at every opportunity. To illustrate the process of genetic transmission through natural selection, Crick drew on the work of the biologist Richard Dawkins, and, via Dawkins, on the work of Shakespeare:

coaxing and complaining to one another. More than three centuries before Freud, Shakespeare had grasped the vagaries of desire and had done so not through the lens of 'theory' but through the instrument of the *theatron*: the 'beholding place,' as his contemporary George Puttenham had defined it in his *Arte of English Poesie* (1589), where men can become women, women become boys, boys become men, and men become gods – or animals.

Thou art translated

Puck's 'translation' of Bottom into a man with an ass's head is the most famous example of transformation in the play, and at first glance it would seem to conform to one of two models that Shakespeare explores: transformations in substance that actually take place through occult or metaphysical processes, and transformations that appear to occur but which are in fact the result of alterations in sense of perception. The effects of the potion on Titania and on the young lovers are the best examples of the latter kind; neither Helena or Hermia actually change but only appear different in the eyes of Demetrius and Lysander: taller or shorter, more lovely or more hateful, more light or more dark-complexioned than they appeared before entering the forest. But when Bottom leaves the stage momentarily, he seems to return as an actual monster:

> SNOUT: O Bottom, thou art changed. What do I see on thee?
> BOTTOM: What do you see? You see an ass-head of your own, do you?
> QUINCE: Bless thee, Bottom, bless thee. Thou are translated.
> (3.1.109–13)

Dawkins gives a very pretty example to refute the idea that natural selection could not produce the complexity we see all around us in nature. The example is a very simple one, but it drives the point home. He considers a short sentence (taken from *Hamlet*):

> METHINKS IT IS LIKE A WEASEL.

He first calculates how exceedingly improbable it is that anyone typing at random . . . would by chance hit on this exact sentence . . . He calls this process 'single-step selection.' He next tries a different approach, which he calls 'cumulative selection.' The computer chooses a *random* sequence of twenty-eight letters. It then makes several copies of this but with a certain chance of making random mistakes in the copying. It next proceeds to select the copy that most resembles the target sentence, however slightly. Using this slightly improved version, it then repeats this process of replication (with mutation) followed by selection. In the book Dawkins gives precise examples of some of the intermediate stages. In one case, after thirty steps, it had produced:

> METHINGS IT ISWLIKE B WECSEL

and after forty-three steps it had the sentence completely correct. *How many* steps it takes to do this is partly a matter of chance . . . The point is that by *cumulative* selection one can reach the target in a relatively small number of steps, whereas in *single-step selection* it would take forever. (*What Mad Pursuit*, 28–9)

Bottom himself feels scratchy about the face and develops a new taste for oats and provender; Titania promises to 'kiss thy fair large ears' (4.1.4) and, upon awaking, finds that he appears fascinating and desirable even in his altered form:

> I pray thee, gentle mortal, sing again.
> Mine ear is much enamoured of thy note;
> So is mine eye enthralled to thy shape;
> And thy fair virtue's force perforce doth move me
> On the first view to say, to swear, I love thee.
>
> (3.1.130–4)

The immediate difficulty with drawing these distinctions in types of transformation, however, is that the only 'objective' evidence for Bottom's change of shape is the 'subjective' perception and reaction of the other characters, and since the primary effect of the plot of the young lovers is to show how unreliable the senses can be, we are left with no absolute criteria for judging how the scene should be played and whether Bottom has actually been transformed or not. Snout's question and Bottom's response allude to this very problem: maybe what Snout 'sees' is in fact his own projection (and his question hangs unanswered).

The crux is finally a theatrical one: does the actor playing Bottom wear a prop or not? The Folio text of the play includes the stage direction 'Enter Piramus with the Asse head,' which would seem to refer to an actual prop. But prop or no prop, the mimetic power of the Elizabethan stage ensures that the joke will be carried off. Titania either strokes the ears of a man wearing an ass's head and provokes our laughter at her misplaced endear-

No doubt both Dawkins's and Crick's choice of Shakespeare to illustrate the problem of the genetic code and the process of natural selection is an overdetermined one, since for so long Shakespeare has represented our most fundamental attitudes about ourselves. More than any other writer, Shakespeare would seem to have grasped the underlying essence that makes us all, as 'humans,' most alike, despite the distance of history, of culture, of gender, and of belief. Ironically, however, we find that both Dawkins's and Crick's use of Shakespeare implies the opposite view: far from regarding Shakespeare as a unique creative intellect, both Dawkins and Crick offer us a 'posthuman' Shakespeare, one who exists only as the effect of inevitable logical probabilities and mathematical patterns and about whom there is nothing remotely human to embrace. Peering into Dawkins's computer, we watch as 'Shakespeare' gradually stabilizes into a dimly recognizable portrait: once *mimesis* has been digitalized, then *Hamlet* is only a matter of processing power.

Shakespeare has never been human

As Crick's own analogies should suggest to us, 'Shakespeare' is indeed our clone and creation, the result of hundreds of years of automated reproduction through printed books, from the very first quartos to the First Folio of 1623 to the many hundreds of editions used today by scholars and students all over the world. If one of the dominant fears about human cloning concerns the notion of 'multiple copies of a human being,' as Wendy Doniger has put it, then a similar fear motivates the scientists at Oxford, Cambridge, Riverside, Norton, Arden, and Pelican, all of whom

ments, or she strokes the ears of a man whom she *believes* to be an ass but who actually looks perfectly human, showing herself to be more perverted than we realized. Bottom either looks like an ass or merely seems asinine – but is there any real difference? After all no one has *actually* been transformed, and Titania is nothing more than an enthralled theatergoer, enraptured by the mimetic arts, and we are as charmed as she is, and as perverted, because all we really want is for Bottoms to be asses, and especially asses who receive such splendid treatment.

Theatrical *mimesis* casts a looming shadow onto the stage, or opens a seam or cranny in its surface, so that we, too, can say, like Hermia after awakening from her night in the forest, that 'Methinks I see these things with parted eye, / When everything seems double' (4.1.188–9). An actor playing a role enters and begins calling the names of men playing roles who are preparing to rehearse a play in which they will play a role with another name. 'Ready. Name what part I am for, and proceed' (1.2.17): Bottom is the most giddy, and like the base 'handicraftman' (4.2.9) and 'mimic' (3.2.19) that he is, he wants as many names and as many roles as possible:

> What is Pyramus? A lover or a tyrant? (1.2.19)
> I will move storms. I will condole in some measure. (1.2.22–3)
> I could play Ercles rarely. (1.2.24–5)
> . . . let me play Thisbe too. I'll speak in a monstrous little voice. (1.2.45–6)
> Let me play the lion too. I will roar that I will do any man's heart good to hear me. I will roar that I will make the Duke say 'Let him roar again; let him roar again.' (1.2.63–6)

work overtime to ensure that the Shakespeare code-script is sold as widely as possible, but only in legitimate copies.

This 'conservative' attitude toward the Shakespearean DNA is arguably more appropriate to a sperm bank than to the fertile seedbed of the seminar, for it ignores the accidents of mutation. Whereas editors used to regard the printed text as a relatively imperfect reproduction of an imagined prior script, most scholars now recognize the many contextual and systemic factors that contribute to the 'Shakespeare' that we read in printed form. The 'source' of the Shakespearean code resides less in the hand of a once living, human person than in a distributed network of many different agents, from actors and other people involved in the production of the plays in the theaters, to the censoring and approving authorities, to the publishers, printers, proof-readers and compositors who actually set the type, and even to the physical printing machine itself – a material or non-human agent – whose 'characteristics' as a particular kind of technology introduced certain parameters on how the text could be assembled and printed and thus on the way that the play actually appeared on the page. Many people are surprised to find that none of Shakespeare's plays survives as a script in his handwriting: modern scholars possess only the printed editions of his plays, many of which record variant words and spellings and even omit entire scenes that we have come to expect.

And given Shakespeare's emblematic status for our most human characteristics, and given our enduring affection for the many seemingly human characters he was able to create, we should not be surprised to find that changing attitudes toward the status of the Shakespearean text are exactly homologous to changing paradigms of the 'gene' in molecular biology over the course of the

I will aggravate my voice so, that I will roar you as gently as
any sucking dove. I will roar you an 'twere any nightengale.
(1.2.74–5)

The bestial pleasures that Bottom embraces are precisely what made the stage at once so enticing and so disturbing to early modern moralists, since, as Erica Fudge has pointed out, the antitheatricalists objected not merely to the perceived threat of erotic perversion and gender transvesticism but also to *species*-tranvesticism, or 'to act the beast,' as William Prynne put it: 'What is this but to obliterate that most glorious image which God himself hath stamped on us, to strip us of our excellency, to prove worse than brutes?' (Fudge, *Animal*, 61). Bottom is so *inside* the role, so fully *become* ass, that human sentence and humanist *sententia* alike are interrupted:

I have had a most rare vision. I have had a dream past the wit
of man to say what dream it was. Man is but an ass if he go
about to expound this dream. Methought I was – there is no
man can tell what. Methought I was, and methought I had –
but man is but a patched fool if he will offer to say what
methought I had. The eye of man hath not heard, the ear of
man hath not seen, man's hand is not able to taste, his tongue
to conceive, nor his heart to report what my dream was. I will
get Peter Quince to write a ballad of this dream. It shall be
called 'Bottom's Dream,' because it hath no bottom, and I
will sing it in the latter end of a play, before the Duke.
(4.1.201–15)

twentieth century. This conceptual adjustment is marked most obviously by the difference between the term 'gene' and the term 'genome': whereas once the 'gene' could be described as a governing agent – as in the common phrase 'gene action,' or in Erwin Schröedinger's notion that the gene is the 'law-code and executive power . . . architect's plan and builder's craft . . . in one' (Keller, *Refiguring Life*, xv, 19, *passim*) – now we understand genes as 'amorphous entities of unclear existence ready to vanish into the genomic or developmental background at any time' (*Concept of the Gene*, x). Rather than a single, executive, Theseus-like 'gene,' the 'genome' describes a dynamic system or 'extensive editorial process' (Keller, *Century*, 67) characterized by a network of relations with other parts of the DNA and other elements of the cell structure. Much of this recursive relationship remains opaque to modern science, and this is partly because of the extraordinary redundancy that seems to be typical of biological information: individual genes may be replaced or mutated, and the cell, for instance, can still develop normally, as other genes or relations among genes compensate for the change.

Subsequent work in the field of genetics has shown that Crick's metaphor for DNA as type, print, and unit of 'exact replication of information' was merely that: a convenient metaphor that actually obscured many aspects of intracellular development. In the words of Keller,

> the [double-helical] structure of DNA provides only the beginning of an explanation for this high fidelity. In fact, left to its own devices, DNA cannot even copy itself: DNA replication will simply not proceed in the absence of the enzymes required to carry out the process. Moreover, DNA is not intrinsically stable:

Since moments like these constantly remind us of Shakespeare's own artifice, we are left with the realization that *we* are the ones who have experienced a kind of *Dream* that is really a clever piece of poetic magic, one in which the most outrageous transformations have occurred only with the help of our own imaginations, the suspension of our normal senses, and the willing credulousness of our own eyes.

The humor of the scenes derives from Shakespeare's uncanny ability to mock the stage at the same time as he takes theater as seriously as possible: the 'rude mechanicals' are bad actors, but the actors playing them are *good* bad actors and thus place the very idea of acting itself in quotation marks, leaving it impossible to tell who is acting and who is not. Critics have often observed that Bottom's speech is a playful travesty of Paul's first letter to the Corinthians, and Shakespeare may intend an oblique jab at the Prynnes and the Gossons: the point is not simply that Bottom has a bad memory for Scripture and thus looks foolish in misquoting it but that the language of the Bible itself often resembles nothing so much as poetry, a strange and beautiful language that speaks of mysteries and that should never be reduced to doctrine or demagoguery. The language of the Bible, Shakespeare suggests, is more properly the language of ecstasy; like myth and like theater, it provides a mode of expression for the unfathomable, for an experience that is beyond the human, for a synesthesia in which the component parts of the body disassociate and begin 'thinking' for themselves by generating new forms and new songs.

its integrity is maintained by a panoply of proteins involved in forestalling or repairing copying mistakes, spontaneous breakage, and other kinds of damage incurred in the process of replication. Without this elaborate system of monitoring, proofreading, and repair, replication might proceed, but it would proceed sloppily, accumulating far too many errors to be consistent with the observed stability of hereditary phenomena – current estimates are that one out of every hundred bases would be copied erroneously. (*Century*, 26–7)

The human genome, it turns out, is less a 'line of computer code' than a garbled sentence packed with thousands of small units that have no clear purpose: structural components and activation points, transposable elements that leap from point to point in unpredictable ways, introns or so-called 'junk' DNA that is spliced out during the reduplication process. A mere 1% of the human genome consists of genes that actually code for the proteins necessary to build new cells.

If the genome consists of an 'elaborate system of monitoring, proof-reading, and repair', then nowhere is the instability of the Shakespearean text more clearly demonstrated than in the phenomenon of Shakespearean 'character,' as the critic and editor Random Cloud-McCleod has shown. From the first quartos of Shakespeare's plays to the Folios printed posthumously to the first collected editions of Nicholas Rowe and Alexander Pope, Cloud-McCleod argues, we find an inevitable tendency toward mutation:

quartos and folios were frequently reprinted, one printing often serving uncritically as copy for the next, the compositors

Of dramatology

It would be difficult to find a better example of Shakespeare's experiment in *mimesis* than the character of Puck, a figure for Proteus, the mythic trickster, and the dramatic fool rolled into one, the sponsor of random events and the superstitions they provoke, of minor social humiliations and the amusement we feel at the expense of others. Puck shifts his own shape so easily that it is difficult to say what his 'true' form really is: to be a 'puck' is to refuse the ontological categories that reason depends upon for its logic and law; it is to revel in chance, accident, and the power of mutation; to become animal, eschewing speech for the raw vocalization of the beast:

> Sometime a horse I'll be, sometime a hound,
> A hog, a headless bear, sometimes a fire,
> And neigh, and bark, and grunt, and roar, and burn,
> Like horse, hound, hog, bear, fire, at every turn.
> (3.1.103–6)

Puck's shape-shifting never settles into any singular form but mutates across a spectrum of non-human beings, pushing beyond even the animal 'itself' to include a primary level of elemental forces. These forces manifest themselves as sounds that are significant but which are not yet names, sounds that can only become names through a metaphorical translation of attributes ('*like* horse, hound, hog, bear, fire'). Through an assertion of will ('I'll') that is simultaneously a moment of self-projection toward an anticipated future state, Puck translates accidents into substance and mere sound into the name for a

modernizing graphic features and punctuation, for example, as they went. Compositors would attempt to correct what seemed like obvious mistakes, but, human nature being what it is, would also create new errors . . . Perhaps the simplest way to characterize this period is to say that Shakespeare's text was drifting. (94)

The early texts show very little consistency in character names and speech-prefixes, even within a single printing of a single play: rather than a single, fixed, unitary character that precedes the code and that we project back as a fully 'human' person (and upon which we base our assessment of Shakespeare's own 'humanity'), the play texts record only a plurality of identities for any one character, who emerges through the code as its effect rather than as it cause. 'Not only is it not philosophically necessary to ascribe a primary or a transcendent unity to the notion of individual, isolated character that so obsesses modern history,' Cloud-McCleod argues, 'but also the text and Shakespeare's nomenclutter [sic] resists such appropriation' (93). One of the most amusing aspects of the 'rude mechanicals' who rehearse 'Pyramus and Thisbe' in the forest outside Athens in *A Midsummer Night's Dream* derives from the fact that the script is so unstable and generates 'characters' who show us nothing but the process of their own mutation. It is as if Shakespeare had anticipated the insights of modern editors and had written the scenes of Bottom and company expressly for them.

Once again, 'Shakespeare' teaches us a surprisingly contemporary lesson. If the philosophy behind the Human Genome Project (presided over by no less a figure than James Watson) and other commercial gene-sequencing projects can be summed up as 'char-

unified type of animal being. Indeed, we may say that the verb 'to be' operates for Puck transitively as well as intransitively, since 'to be something' is not simply an act of self-existence but an act of performative creation.

At one level Puck's speech is entirely figurative, since Puck is a symbol for the principle of transformation in general, whether by causes physical or metaphysical, patent or occult. In Agrippa's terms he 'can do all things upon all things' and infuses himself throughout the hierarchy of being. And yet at the same time the speech is entirely literal, since as an actor in a theatrical performance 'Puck' can indeed become any creature and any natural force imaginable. We have reached a magical point where the play-becomes-*play*: not 'drama,' the bounded unity of the critic with his 'plot,' his 'structure,' and his 'act and scene,' but a moment when all that happens does so not because of what has already happened or because of what will happen next; a 'moment' – we can speak only metaphorically of this middle of *Midsummer*, approaching duration but before, after and outside duration as discrete and bounded unity of time – that could go in any direction whatsoever, poised at the edge of the pure act; a 'scene' that unfolds for a time during which we have forgotten the beginning and how we got here and when we still do not know what the end will be and do not care.

This point of pure pleasure and absorption marks not simply the absorption of the audience watching the play but the absorption of the actor. For is it not the special privilege of the actor to be in this magic time for the entire period of the performance, and not just for a moment? Which raises the even larger question of for whom, after all, the play can be said to be: what if the entire performance is not for us, and not for

acter is a coded sequence,' then the Shakespearean text shows us that this 'character' is multiple, contingent, and always changing, more a clone than a unique signature, more cyborg than flesh-and-blood. Actors often refer to characters as though they were stable creatures, but in fact they are virtual creatures who constantly evolve, uncanny simulations in which we recognize, however distantly, some thread of connection to ourselves. Even if we were to accept the notion of a stable 'code' or 'script' – and anyone who studies either Shakespearean play texts or the human genome knows we cannot – the 'character' who results is, after all, completely different at every occasion of textual encounter: every performance, every actor, every reading and projected imagination of the character by every different reader produces not a human 'character' but something more similar to a clone. The actual science of cloning, too, produces not a flawless copy of an original entity (popular fantasies and fears notwithstanding) but only a duplicate of genetic information; the expression of this information will vary according to an infinite number of factors, some 'external' and some 'internal' to our physical development, but all in an important sense a function of context. Clusters of cells in turn always develop in slightly different ways, so that even identical twins raised in the same womb display a distinct somatic shape.

But in their persistent mistakes and creative reappropriation of the script, in their eagerness to play all roles at once and to embrace the codes and artifice necessary to project 'character' in as convincing as way as possible, Bottom the weaver and his company raise a further enormous question: perhaps modern metaphors for life as script, information, and code do not derive from the computer at all – perhaps they derive from the *theater* and from performance. The critic Richard Doyle has pointed out that:

Theseus; what if the shuffling, unintentional comedy of the rude mechanicals and Puck's apology to the theater audience at the close of the play is all a ruse or subterfuge, a camouflage under which the true performance takes place for the actors themselves, and not for any audience? What if the performance of Puck is for becoming-Puck, and nothing more: for the high of the role as potion, but for real: to be in the state of mimetic hallucination?

Rude mechanicals

Thus when Bottom the weaver, Quince the carpenter, Flute the bellows-mender, Snout the tinker, Snug the joiner, and Starvling the tailor enter the stage of *A Midsummer Night's Dream* to rehearse their own play-within-the play, it is tempting to imagine that Shakespeare the actor was among them, playing the role of Puck: the scenes offer his amused recognition of how close poets, actors, and artisans remained at the close of the sixteenth century, in social status, economic organization, and in working methods. James Burbage, father of Richard, the most famous actor in Shakespeare's company, had himself been both an actor and a joiner before building the Theatre, where *A Midsummer Night's Dream* was probably first performed. Ben Jonson was a liveried member of the Tilers and Bricklayers Company; Shakespeare himself was the son of a glover. Indeed, when Puck 'translates' Bottom into a man with an ass's head, he employs a glover's term of art: to 'translate' leather or cloth was to cut and shape it into a new form, often by using leftover pieces. But Puck also uses the *poet's* tool: *translatio* was the technical Latin word for metaphor (the translation

the notion that life is a sequence of instructions, rather than produced by the invention of the computer, is actually rhetorically feasible before the widespread notion of the computer program. Indeed, what made the equation between vitality and information possible was a shift not just in the technology of information, but in the articulation of life. ('Emergent Power,' 310)

Does not this new articulation of life derive as much from the theater and the book as from the automata and animal-machines hypothesized by Descartes or the late-medieval astrological clocks that modelled the motions of the universe, two famous examples of early modern technologies of information cited by Doyle? And today's technologies push the horizon even further, as Timothy Lenoir has argued:

Our posthuman era is characterized by a seamless articulation of intelligent machines and human being . . . We need to explore not only the content of inscription (philosophical debates and abstract ideologies, along with representations of the posthuman in fiction, film, videogames, and other media), but also the technologies of inscription: computer code, object-oriented languages, algorithms, chips, 'smart' fabrics, nanodevices, radio frequency ID tags, and molecular assemblages – all the means for incorporating representations of the posthuman and improvising new repertoires of physical gesture and performance. We need to understand how the posthuman 'gets under our skin,' to use Bernadette Wegenstein's metaphor – both figuratively, in the sense of images and advertising campaigns attempting to shape our cultural

of the Greek term, we could even say), which Aristotle had defined as 'the movement of an alien name from either genus to species or from species to genus or from species to species or by analogy' (*Poetics* 21 1457b). To use metaphor is not simply to 'move,' to 'transfer,' to engage in 'transference,' to 'turn,' or even to 'pervert' a thing from its proper object: it is to create a chimeric substance by grafting one species of thing with another one alien to it.

This was precisely the argument of George Puttenham in his *Arte of English Poesie* (1589): since the poet alters and even improves on nature, he is best compared to a doctor who administers medicine, a lens-maker who grinds spectacles, a gardener who cross-breeds plants, and an alchemist who fabricates gold (310). Moreover, the poet accomplishes his task using knowledge 'gathered by experience' (21), and for this reason his poet's knowledge is both 'mechanicall' and 'scientificke' – the term is Puttenham's (19): he is like a carpenter, a joiner, a tailor, or a smith, since he 'makes things and produceth effects altogether strange and diuerse, and of such forme and quality (nature always supplying stuffe) as she neuer would nor could haue done of her selfe' (310).

But if nature operates like a craftsman or poet, and the poet and the craftsman both imitate natural techniques when they accomplish the products of their art, then how can we make any meaningful distinction between the 'natural' and the 'artificial,' whether in terms of substance or of process or of purpose? Throughout the play, the faeries are the personification (if the term can have any meaning in this context) of a power that produces naturally artificial and artificially natural hybrids, using things that share some quality while remaining different

imaginary, and literally, in the interfaces of machinic relays with the body. ('Makeover,' 212)

Doyle has even described the new horizons of the laboratory as 'postvital' and not merely as 'posthuman,' since developments in nanotechnology, computing, robotics, cybernetics, and systems theory have begun to blur our most fundamental distinctions between living and non-living entities. Are the 'rude mechanicals' 'alive'? The question becomes more difficult to answer the more you think about it. Rodney Brooks, an early pioneer in the field of artificial intelligence, describes the robot, or 'situated creature,' as:

one that is embedded in the world, and which does not deal with abstract descriptions, but through its sensors with the here and now of the world, which directly influences the behavior of the creature. An embodied creature or robot is one that has a physical body and experiences the world, at least in part, directly through the influence of the world on that body. A more specialized type of embodiment occurs when the full extent of the creature is contained within that body. (51–2)

We could find no better definition of the dramatic character. If Bottom and company tell us anything, they tell us that Shakespeare's theater was the first modern cybernetic system, a machine in which the poet grafted the living body of the actor to the props and prosthetics that fabricated personhood. Perhaps we should regard the Shakespearean character as the original robot – the human has become posthuman before it even became human in the first place!

in species. Titania sleeps on a bed of flowers, Oberon informs Puck:

> I know a bank where the wild thyme blows,
> Where oxlips and the nodding violet grows,
> Quite overcanopied with luscious woodbine,
> With sweet musk-roses, and with eglantine.
> There sleeps Titania sometime of the night,
> Lulled in these flowers with dances and delight . . .
> (2.1.249–54)

One moment the plants grow naturally, the next they burst into a 'canopy' over the natural bed; the transformation occurs through Oberon's poetic language, as metaphor works its mysterious operation at the center of the speech to form a new form, part nature, part art, the cascade of rhyming couplets imitating the entanglement of the woodbine and the eglantine with one another. Titania repeats a similar image later in the play, when she seduces Botttom-as-ass:

> So doth the woodbine and the sweet honeysuckle
> Gently entwist; the female ivy so
> Enrings the barky fingers of the elm.
> (4.1.41–3)

The play's editors gloss 'woodbine' as 'bindweed or convolvulus' and observe that 'the honeysuckle . . . always twines in a left-handed helix' while 'the bindweed family . . . always twines in a right-handed helix' (citing Martin Gardner's *The Ambidextrous Universe* [1970], 62), noting the 'mixed up violent' nature of the

We have never been human, either

Today, the idea of life as code and nature as a book survives in a miraculous but peculiarly literal object: the DNA Book Series, published by the Japanese RIKEN Corporation and available in six separate titles, including *Rice*, *Mouse*, and – inevitably – *Human*. Each physical book, paperbound and conveniently stored on a shelf, serves as a distribution vehicle for clones of DNA sequences; the actual genetic material arrives 'stained' to the pages of the book on bright red dots, which may be cut out with scissors and dissolved in solution for experimental purposes – always using gloves, of course, and always keeping the pages dry, as the book's prefatory materials expressly indicate. Now we can truly say, with Milton, that 'books are not absolutely dead things, but do contain a potency of life in them to be as active as that soul was whose progeny they are; nay, they do preserve as in a vial the purest efficacy and extraction of that living intellect that bred them' (*Areopagitica*, 720.1). And so they do: the very book you are currently holding has, after all, become a personalized 'book of life,' with genetic material from your fingertips, hair, clothing, and sneezes smudged across its pages, preserved for any scientist with the appropriate laboratory, growth-factor, and inclination to regenerate a version of yourself.

This self would be you, and it would also be not you. And you would be human, but you would also not be. For DNA is certainly not a universal language common to all humanity, as Jason Robert and Françoise Baylis observe:

> Although human beings might share 99.9% commonality at the genetic level, there is nothing as yet identifiable as

image rather than the pleasurable eroticism implied by 'sweet,' 'gently' and the image of the ring.

Oberon's subsequent lines employ the principle of hybrid productivity to even more spectacular effect by applying the techniques of the humble glover's workshop to exotic materials: 'there the snake throws her enameled skin, / Weed wide enough to wrap a fairy in' (2.1.255–6). While walking in the forest, we may have once paused under a natural canopy. We may once have worn snakeskin boots or belts. But we have never wrapped our entire bodies in a snake skin that lies hidden in the grass. A related image recurs a few lines later, when Titania orders several faeries to make 'war with remrice [bats] for their leathern wings / To make my small elves coats' (2.2.4–5); again the lines emphasize a change of shape effected through creative translation, as bat-wings are refashioned into a leather-like substance and stitched by the operation of metaphor into a little coat. Soon Titania commands the elves to steal honeybags from bees and to 'crop their waxen thighs' and use them for candles ('night tapers'), which they will 'light ... at the fiery glow-worms' eyes'; and to 'pluck the wings from painted butterflies / To fan the moonbeams from his sleeping eyes' (3.1.159–64). In each case, the faeries stretch the notion of 'purpose' itself to its limit, crossing recognizable human products (candles, fans) with substances that are so inhuman that they remind us how alien and purposeless the so-called 'art of nature' really is. For the notion of 'purpose' is, strictly speaking, a human notion that we project onto the natural world in order to find it charming (or tragic, or otherwise significant). But the world is not our mirror, as the philosopher Giorgio Agamben has pointed out in his genealogy of philosophy's engagement with what it persistently refers to, in a homogenizing singular, as 'the animal':

absolutely common to all human beings. According to current biology, there is no genetic lowest common denominator, no genetic essence, no single, standard, 'normal' DNA sequence that we all share. Moreover, comparative genomic research has thus far been of no help in establishing the boundary of human species identity. Much of 'our' DNA is shared with a huge variety of apparently distantly related creatures (e.g., yeast, worms, mice). Indeed, given the evidence that all living things share a common ancestor, there is little (if any) uniquely human DNA. More strikingly perhaps, though human beings are morphologically and behaviorally vastly different from chimpanzees, we differ genomically from chimps by no more than 1.2–1.6%. (4; citing Lewontin, 36)

How, given these findings, can we continue to defend a sovereign notion of the human, or even of 'species' in general? 'At present, there are somewhere between nine and twenty-two definitions of species in the biological literature,' Robert and Baylis point out: 'of these, there is no one species concept that is universally compelling' (3).

Today the fields of science and bioethics provide our language for discussing the problems presented by the chimera: but in Shakespeare's period that language was provided by poetics and literary criticism. His young lovers dream of snakes chewing at their hearts and tell of a miraculous potion fabricated from a rare flower and in need of a test upon a human subject, all while a chimeric man-donkey tries to breed with a Fairy Queen and brays and laughs simultaneously during his seduction. Many Renaissance writers embraced the idea of mixing categories and blending natural and artificial processes, pointing out that many

> Too often . . . we imagine that the relations a certain animal subject has to the things in its environment take place in the same space and in the same time as those which bind us to the objects in the human world. This illusion rests on the belief in a single world in which all living beings are situated. Uexküll ['Jakob von Uexküll, today considered one of the greatest zoologists of the twentieth century and among the founders of ecology'] shows that such a unitary world does not exist, just as a space and a time that are equal for all living things do not exist. The fly, the dragonfly, and the bee that we observe flying next to us on a sunny day do not move in the same world as the one in which we observe them, nor do they share with us – or with each other – the same time and the same space. (*The Open*, 40)

This ontological heterogeneity and distance is accentuated in *A Midsummer Night's Dream* through Shakespeare's fascination with smallness, one of the play's most delightful and memorable effects: 'Peaseblossom, Cobweb, Mote, and Mustardseed' (3.1.153) are emblematic of the insubstantial, the minute, the atomic, and even the nanotechnological (which we could describe as the goal-directed modification of natural objects at a small scale).

Metaphormosis

Just as when we look through a microscope, however, to find that minute things have grown enormous, so also we realize that the tiny faeries are vehicles for forces of growth, decay, and metamorphosis that extend throughout the entire scale of the

commonplace objects already did so. And no one expressed the paradoxes that resulted more beautifully than Shakespeare himself in a famous passage of *The Winter's Tale*:

> PERDITA: For I have heard it said
> There is an art which, in their piedness, shares
> With great creating nature.
> POLIXENES: Say there be;
> Yet nature is made better by no mean
> But nature makes that mean: so, over that art,
> Which you say adds to nature, is an art
> That nature makes. You see, sweet maid, we marry
> A gentler scion to the wildest stock,
> And make conceive a bark of baser kind
> By bud of nobler race. This is an art
> Which does mend nature – change it rather – but
> The art itself is nature.
> PERDITA: So it is.
> POLIXENES: Then make your garden rich in gillyvors,
> And do not call them bastards.
>
> (4.4.86–99)

As the lines allow nature and art to intertwine, they seem to suggest that drawing an absolute line between them may be impossible if not irrelevant, especially when confronted with the delicate beauty of a streaked flower.

Polixenes's speech reveals one of the most surprising aspects of Renaissance theories about art and nature: the claims made for the poet at the turn of the seventeenth century are remarkably similar to the debates over biotechnology and its

natural world. These are the forces of the seasons and the weather, as we have seen, but also of the pasture, the valley, and the garden:

> Over hill, over dale
> Thorough bush, thorough briar,
> Over park, over pale,
> Thorough flood, thorough fire
> (1.2.2–5)

The faeries deposit 'dew' in 'orbs upon the green' (2.1.9) and ornament the flowers with 'spots' that seem like 'rubies,' 'freckles,' and 'pearls' (2.1.11–15); they 'creep into acorn cups' (2.1.31) and 'kill cankers in the musk-rose buds' (2.2.3). This same transformative and sometimes violent power touches the human mortals in the play, since much of the comedy in the young lovers' plot results from the way that Shakespeare amplifies the transformative effects of the potion with the splicing tool of metaphor – we may now call it a technique not of metamorphosis but of *metaphormosis*. Hermia becomes a 'cat,' a 'burr,' a 'vile thing,' a 'serpent,' a 'dwarf,' a '*minimus* of knot-grass' weed, a 'bead,' an 'acorn,' even a 'Tartar' and an 'Ethiope' (two species of human non-humans in Shakespeare's period, we could say). Helena, meanwhile, becomes a 'goddess, nymph, perfect, divine!' (3.2.137), her lips crossed with ripe fruit ('those kissing cherries'), her body anatomized by the lover's x-ray eye:

> Transparent Helena, nature shows art
> That through thy bosom makes me see thy heart
> (2.2.109–11)

potential at the turn of the twenty-first, especially the debates over normative definitions of identity, personhood, and the human. Sir Philip Sidney's famous *Defence of Poesy*, written only a few years before *A Midsummer Night's Dream*, invokes traditional arguments about *mimesis* and the artifice of poetry, and then suddenly leaps far beyond them. All 'arts,' Sidney argues, using a theatrical metaphor, imitate natural processes and natural things, 'without which they could not consist, and on which they so depend, as they become actors and players, as it were, of what nature will have set forth' (99.37–100.2). This interest in nature, Sidney maintains, unites the astronomer's study of the stars, the geometrician's and arithmetician's analysis of quantity, the musician's principles of harmony, the natural philosopher's inquiry into the causes of organic growth, generation, change, and decay, the moral philosopher's precepts for regulating behavior, the accounts of human nature offered by the lawyer and the historian, the 'rules of speech,' persuasion, and argument provided by the grammarian, the rhetorician, and the logician, the physician's prescriptions for our bodies, and the metaphysician's intellectual speculation into 'abstract notions' that 'build upon the depth of nature' (100.2–20).

But 'of all sciences,' Sidney declares, 'is our poet the monarch' (113.18–20), and this is because only the poet goes *beyond* nature, using the force of his creative intellect to achieve power over nature with no limitation:

> Only the poet . . . lifted up with the vigour of his own invention, doth grow in effect into another nature, in making things either better than Nature bringeth forth, or, quite anew, forms

Of all the characters, Helena speaks in the most 'poetic' language of the play – the densest, most syntactically complex, the most vivid, the most elaborately formal – and she undergoes the greatest metaphormoses. Driven by jealousy and obsession, she trails whining after Demetrius as he searches the wood for the fleeing lovers:

> I am your spaniel, and, Demetrius
> The more you beat me I will fawn on you.
> Use me but as your spaniel: spurn me, strike me,
> Neglect me, lose me; only give me leave,
> Unworthy as I am, to follow you.
> What worser place can I beg in your love –
> And yet a place of high respect with me –
> Than to be used as you use your dog?
>
> (2.1.203–10)

If, as Aristotle argues, one of the most effective ways to denigrate someone is to use a metaphor borrowed from the same genus but of a worse type – like 'saying . . . that a person praying "begs," because both are forms of asking' (*Rhetoric* 3 1405a), much the way Helena 'begs' Demetrius here, or the way Titania is made to 'beg' Oberon's patience (4.1.57) – then, we may ask (at the risk of stretching Aristotle's argument beyond itself), how much more degrading is a metaphor drawn from an entirely *different* genus and species? Although Helena's lines are often played for humor in performance, they suggest a self-negation that approaches the pathological; in the context of her earlier declarations ('herein I mean to enrich my pain' [1.1.250]), 'I am your spaniel' rings as a statement of being and

such as never were in Nature, as the Heroes, Demigods, Cyclops, Chimeras, Furies, and such like . . . Nature never set forth the earth in so rich tapestry as diverse poets have done; neither with pleasant rivers, fruitful trees, sweet-smelling flowers, nor whatsoever else may make the too much loved earth more lovely. Her world is brazen, the poets only deliver a golden. (100.21–33)

The ecological overtones to Sidney's arguments are striking and deserve further study, since the promise of poesy seems to include a new appreciation for the complexity of organic systems as well as a vision of intervening and altering those systems, perhaps for human benefit. But Sidney has even grander claims in mind: the poet's greatest accomplishment, he argues boldly, is nothing less than the perfection of the human itself. Some poets make chimeras or monstrous hybrids; others give birth to supermen. Theagenes, Pylades, Orlando, Cyrus, Aeneas: heroes such as these are better than living people, even though they resemble them in every respect. Nor can we dismiss these creations as mere 'imitation or fiction' (101.2), Sidney maintains, immediately anticipating the later objections of a Bacon or a Galileo, both of whom dismissed the poet as a maker of mere 'chimeras.' For in fashioning his creatures, the poet-as-maker draws upon an idea of the human that is more real and more vital than the two-legged creatures who walk the earth. 'Any understanding knoweth the skill of the artificer standeth in that *Idea* or fore-conceit of the work and not in the work itself' (101.2–5), Sidney claims, warming to his subject:

not merely as a figure of speech. As the self-loathing blossoms, so too do the animal figures: she is 'as ugly as a bear, / For beasts that meet me run away for fear' (2.2.100–1). Soon Shakespeare coaxes Helena's mutations toward the mineral and the machine:

> You draw me, you hard-hearted adamant,
> But yet you draw not iron, for my heart
> Is true as steel.
>
> (2.1.195–7)

As men become magnets and organs fossilize into stone, we are left with the image of a fervent young woman whose metal heart beats with the steadfast devotion of a mechanical implant – a cyborg manufactured through the grafting tool of metaphor, but the cyborg, too, as a figure *for* desire, which is itself figured through the occult force of magnetism.

Drop by drop into Helena's soliloquies, metaphor dissociates qualities and recombines them in unexpected ways; with a sudden burst, her speech buckles, twists around itself, and unravels in a helix of pure poetry:

> Call you me fair? That 'fair' again unsay.
> Demetrius loves your fair – O happy fair!
> Your eyes are lodestars, and your tongue's sweet air
> More tuneable than lark to shepherd's ear,
> When wheat is green, when hawthorne buds appear.
> Sickness is catching. O, were favour so!
> Your words I catch, fair Hermia; ere I go,
> My ear should catch your voice, my eye your eye,
> My tongue should catch your tongue's sweet melody.

Which delivering forth also is not wholly imaginative, as we are wont to say by them that build castles in the air; but so far substantially it worketh, not only to make a Cyrus, which had been but a particular excellency as Nature might have done, but to bestow a Cyrus upon the world to make many Cyruses, if they will learn aright why and how that maker made him. Neither let it be deemed too saucy a comparison to balance the highest point of man's wit with the efficacy of Nature; but rather give right honour to the heavenly Maker of that maker, who having made man to His own likeness, set him beyond and over all the works of that second nature: which in nothing he showeth so much as in Poetry, when with the force of a divine breath he bringeth things forth surpassing her doings . . . But these arguments will by few be understood, and by fewer granted. (101.7–25)

How, indeed, can we grant Sidney's astonishing claims? He has just claimed that poetic 'imitations' or 'fictions' – derived from the Latin *fingere*, to fashion or to form – are as real and 'substantial' as any natural body, while at the same time comparing the process of poetic creation to the first act of God himself. By studying nature and adapting its principles to go beyond what nature makes, the poet pays homage to natural processes and divine power simultaneously, and he does so by means of the very processes that God himself has modelled in his own divine poetry!

> Were the world mine, Demetrius being bated,
> The rest I'd give to be to you translated.
>
> (1.1.181–91)

Catalyzed by a viral-like metaphor, adjectives are transformed into nouns, accident into substance, word into thing into concept; body parts separate and float free in the solution of language, wrapped in a syllable, a glance, a kiss, as one entity gradually transforms itself into another. Helena's speech is also the first of several in the play to figure desire as a process of splicing or transplanting of organs – the eye, the tongue, the heart – an image that returns in her impassioned speech to Hermia toward the middle of the play:

> We, Hermia, like two artificial gods
> Have with our needles created both one flower,
> Both on one sampler, sitting on one cushion,
> Both warbling of one song, both in one key,
> As if our hands, our sides, voices, and minds
> Had been incorporate. So we grew together,
> Like to a double cherry: seeming parted,
> But yet an union in partition,
> Two lovely berries moulded on one stem.
> So, with two seeming bodies but one heart
>
> (3.2.203–12)

If ever theater were like a laboratory and poesy a kind of experimental science, then we find it here: Helena and Hermia, grown from the same substance, flesh or 'stem,' arising out of a vitalizing power that lives in single parts (hands, sides, voices) and that

The forms of things unknown

No doubt we have never been human, except in our fantasy and desire, and despite the bristling number of laws and regulations that would defend an idea of the human in the face of the new entities created in laboratories across the globe. Enter these labs, and we will find human ears grown on the backs of mice for use in plastic surgery and human kidneys grown in mouse abdomens to study xenotransplantation – fully functioning, tiny human organs derived from fetal stem cell transplants that produce urine and that are nourished by the mouse blood of mouse veins. We will find human stem cells in the brains of mice, too, and in the brains of monkeys, and in fertilized rabbits' eggs and chick's eggs. We will find the parts for an entirely new zoology of life, and the blueprints for its industrial uses: the teeth of pigs sewn into rat intestines, pigs circulating human blood, pigs growing organs modified to enhance the prospect of donation to human bodies, pigs generating human insulin for diabetics, goats whose milk contains spider proteins for silk stronger than steel and compatible with human tissue, cows that produce human milk, rabbits that produce milk to rescue babies with severe muscular disease, fluorescent rabbits, monkeys with glowing jellyfish genes, cloned mules – one of the world's oldest chimeras made into a simulacrum of itself. Genetically modified moths have been released in Arizona to fight the pink bollworm, one of the most successful cotton parasites. 'Our ultimate plans are to insert conditional lethal genes that will fight against this enormously successful tendency to survive and infest cotton,' said Thomas Miller of the Department of Entomology, University of California-Riverside, a modern-day Titania killing 'cankers in the musk-rose

assembles them into a coherent 'person.' Helena's speech registers both Shakespeare's persistent interest in the body as a source of poetic form and figure and at the same time his attempt to push the notion of an integral 'body' to its limit, carving it into its component elements and showing how artificial the so-called natural body really is.

Many of the difficulties raised by modern stem cell research and posthuman entities such as the clone, the chimera, and the xenotransplant emerge here in the image of the twin and the lover. What guarantees the singularity of identity, whether of personhood or of body or of species? Physical integrity? Memory and history? A unique signature, script, or code? Does 'life' reside in parts or in wholes? Internally in discrete organs and cells or externally in the relation between living entities? For some, interrupting the putative propriety and uniqueness of the living entity seems to violate the integrity of the self or of the species category; for others, transplantation between individual bodies, whether of individual organs or isolated genetic sequences, seems perfectly legitimate, since chemical and molecular structures are widely shared across species. As Agamben has argued, such distinctions drive across the category of the 'human' and fissure it from within:

> The division of life into vegetal and relational, organic and animal, animal and human, therefore passes first of all as a mobile border within living man, and without this intimate caesura the very decision of what is human and what is not would probably not be possible. It is possible to oppose man to other living things, and at the same time to organize the complex ... economy of relations between men and animals,

buds' (2.2.3). Soon goats may produce a malaria vaccine and eggs may be infused with cancer-fighting proteins. Allergy-proof cats are now available for purchase, following upon the first cloned house-pet. Llama antibodies may soon be available on a dipstick to determine whether that cup of coffee in your hand really *is* decaffeinated ('There have been several news reports of mix-ups at particular coffee houses,' one project team member explained helpfully).

This is now – this *was* now, since many of these examples are several years old, and the now of biotechnology changes far more rapidly than the now of Shakespeare. What are we to make of these 'fictions,' in the truest etymological sense of the term, these creatures that 'never were in Nature, as the Heroes, Demigods, Cyclops, Chimeras, Furies,' as Sidney described them? They will need new names: the geep, the humanzee; CopyCat (known affectionately at the lab as CC) and ANDi ('inserted DNA' read backwards), the genetically modified monkey. We will need new concepts, since the old ones, even the most basic – 'species,' 'animal,' 'human,' 'nature,' 'life' – will no longer suffice to make sense of the becoming world. And, like Socrates, we will need new questions. How will we reconcile our ethical obligations to these new non-human 'mortals' with the imperative to recognize the rights and freedoms that the category of the 'human' has traditionally enjoyed – rights, indeed, that seem alarmingly to be eroding before our eyes in many parts of the world? What defines the 'human' in the first place, and what is at stake in maintaining that definition? And what, more pointedly, is at stake in our definitions of 'life' and the 'living'? If the 'human' has a future, no doubt it will be as a strategic name, a contingent category to protect against suffering and atrocity but which at

only because something like an animal life has been separated within man, only because his distance and proximity to the animal [sic] have been measured and recognized first of all in the closest and most intimate place. But if this is true, if the caesura between the human and the animal passes first of all within man, then it is the very question of man – and of 'humanism' – that must be posed in a new way. (*The Open*, 15–16)

More than any other play written during the period when humanism was in its moment of ascendance, *A Midsummer Night's Dream* provokes a radical re-questioning of that humanism and the categories of life upon which it depended by splicing across as many categories as possible: human and inhuman, natural and artificial, male and female, small and large, organic and inorganic. Helena is Shakespeare, stitching an entire artificial world into her 'sampler,' mixing species at will, grafting one property to another to produce a forest of new creatures, but she is also Shakespeare's hybrid creation in that world, the actor's living (male) body grafted to an artificial (female) persona and pumped full of *mimesis*: there is no blood, after all, in the veins of a character, no oxygen, only fiction and the imitation of a heartbeat.

Shakespeare has never been modern

In this regard, it would be most accurate to speak of *A Midsummer Night's Dream* not as 'early modern' but as 'never been modern,' to borrow a phrase from the sociologist of science Bruno Latour, who has described as the 'modernist settlement'

the same time does not retain a monopoly on concepts of dignity and 'rights,' a fully political category in a new notion of 'politics' that can accommodate many different entities, living and non-living alike. To speak of a 'posthuman' era is certainly not to deny the many differences between you and me and other forms of non-human entities, since to deny these differences would be asinine, as the philosopher Jacques Derrida has pointed out. Indeed, there is *only* difference, and one task of a posthuman future will be to recognize that thought itself marks a specific difference between human and non-human entities without at the same time resorting to a reflexive notion of human exceptionalism. If we have an obligation to think anew the differences between humans and non-humans (not to mention the differences among humans themselves), we must also recognize that in the case of 'the animal' the thought of difference is impossible: for the moment of difference would be precisely where what we call 'thought' is not. The posthuman would try to think this impossible thought of itself otherwise, which is to say that it will no longer think: it will imagine.

Who can say whether the task of imagining this posthuman future is better suited to the poet or to the scientist? As Francis Bacon described poesy in his *Advancement of Learning*:

> Poesy is a part of learning in measure of words for the most part restrained, but in all other points extremely licensed, and doth truly refer to the imagination; which, being not tied to the laws of matter, may at pleasure join that which nature hath severed, and sever that which nature hath joined; and so make unlawful matches and divorces of things. (86)

the emergence of a worldview that depends on false premises: one in which bodies are regarded as separate from minds, subjects have been distinguished from objects, words from things, human from non-human entities. The more we examine such categorical distinctions, Latour argues, the more we realize how inaccurate they really are, as well as how much effort – institutional, financial, ideological – is necessary to maintain them. And ironically enough, the more we reinforce our categorical distinctions, the more we shall find hybrids proliferating under our noses, if only we know where to look.

This modernist settlement has its origins long before the turn of the seventeenth century, as scholars such as Erica Fudge, Bruce Boehrer, and Laurie Shannon have argued. But a major component of this 'settlement' comes sharply into view around 1600: the distinction between 'literary' and 'scientific' ways of understanding the world. This distinction in turn implies two very different attitudes toward language. One position views language as a creative tool, a metaphorical process that postulates the world with every utterance; the second regards language as a neutral, reflective medium that mirrors the world like a map or a diagram. A version of this modernist settlement is spoken in *A Midsummer Night's Dream* by Theseus, who would like to isolate some uses of language as nonsensical while reserving for himself the right to utterances that can classify the world in concrete and substantial ways. Illuminated by the reassuring light of morning, Theseus dismisses the young people's account of their night in the forest as mere delusion:

> I never may believe
> These antique fables, nor these fairy toys.

Whatever Bacon's own cautions as to the indulgences of poesy – 'rather a pleasure or play of imagination, than a work or duty thereof' (125) – the 'modern' scientific method he announced for himself was precisely one of mixing and conjoining categories that normally seemed ontologically distinct from one another:

> But if anyone conceive that my forms too are of a somewhat abstract nature, because they mix and combine things heterogeneous . . . if anyone, I say, be of this opinion, he may be assured that his mind is held in captivity by custom, by the gross appearance of things, and by men's opinions. For it is most certain . . . that the power of man cannot possibly be emancipated and freed from the common course of nature, and expanded and exalted to new efficients and new modes of operation, except by the revelation and discovery of forms of this kind. (*New Organon*, 153)

This revelation and discovery, the emancipation and freedom from the common course of nature of which Bacon speaks, is a task for anyone 'who asks questions,' as Socrates calls the final authority over names. And we will find these questions not only in the Petri dishes of the laboratory but in our poetry and plays and novels, indeed in all of the teeming culture that surrounds us like a nutrient-rich growth medium. If 'all life evolves by the differential survival of replicating entities,' Richard Dawkins has famously argued, then:

> the gene, the DNA molecule, happens to be the replicating entity that prevails on our own planet. There may be others . . . But do we have to go to distant worlds to find other kinds

Lovers and madmen have such seething brains,
Such shaping fantasies, that apprehend
More than cool reason ever comprehends.
. . .
The poet's eye, in a fine frenzy rolling,
Doth glance from heaven to earth, from earth to heaven,
And as imagination bodies forth
The forms of things unknown, the poet's pen
Turns them to shapes, and gives to airy nothing
A local habitation and a name.

(5.1.2–22)

Theseus bespeaks the intolerance of the sixteenth-century Church and State when confronted with the theaters that crowded their cities and the poets whose 'inventions' threatened to disrupt traditional hierarchies and categories of thought. Like Socrates in the *Republic*, Theseus would ban poetic hybrids from Athens because autocracy tolerates no partner and absolute power comprehends no mixture. But in the woods outside Athens, words take on a life of their own, as the critic Murray Krieger has written of Shakespeare's verse:

> A word seems to turn into another word. It is very exciting to watch it happen. But how can the transformation occur? Here is the word in the process of overrunning its bounds, destroying its own sense of territorial integrity along with its neighbor's. It is undoing the very notion of 'property' . . . so that it is defying the operational procedures of logic and law – and those as well of language itself . . . Because words, however different in meaning, sound alike – or almost alike –

of replicator and other, consequent, kind of evolution? I think that a new kind of replicator has recently emerged on this very planet. It is staring us in the face. It is still in its infancy, still drifting clumsily about in its primeval soup, but already it is achieving evolutionary change at a rate that leaves the old gene panting far behind. The new soup is the soup of human culture. We need a name for the new replicator, a noun that conveys the idea of a unit of cultural transmission, or a unit of *imitation*. 'Mimeme' comes from a suitable Greek root, but I want a monosyllable that sounds a bit like 'gene'. I hope my classicist friends will forgive me if I abbreviate mimeme to *meme* . . . It should be pronounced to rhyme with 'cream'. Examples of memes are tunes, ideas, catch-phrases, clothes fashions, ways of making pots or of building arches. Just as genes propagate themselves by leaping from body to body via sperms or eggs, so memes propagate themselves in the meme pool by leaping from brain to brain via a process which, in the broad sense, can be called imitation . . . As my colleague N. K. Humphrey neatly summed up [the concept of meme]: '. . . memes should be regarded as living structures, not just metaphorically but technically. When you plant a fertile meme in my mind you literally parasitize my brain, turning it into a vehicle for the meme's propagation in just the way that a virus may parasitize the genetic mechanism of a host cell.' (*Selfish*, 192; emphasis in original)

Memes live in literature, but they also live in the modern 'myths,' as Roland Barthes has called them, of our media culture: in advertising, in films; on television, billboards and cell phones; stitched into clothing, pasted on doorways, launched into space –

they are forced, as we hear by watching, to become alike. Often it occurs in the coupling act of rhyming or – more extremely – of punning. But often the poet slips from word to word and from sound to sound in a continuing parade of subtle echoes . . . Shakespeare everywhere reminds us of the transformational power of words, their appearing to defy their own distinctness by overlapping and changing places with one another. (110–12)

Under Krieger's keen eye, Shakespearean language resembles nothing so much as a teeming Petri dish, one in which words have become autonomous entities that grow and change by metabolizing the associated concepts and phonic qualities around them. Regard *A Midsummer Night's Dream* through the 'frenzied' eye of the poet and we see that the challenges posed by *poiesis* at the beginning of the seventeenth century have become those of modern science, and especially of biotechnology, at the opening of the twenty-first. Theseus occupies the White House, the courtroom, and the Senate Chamber, but Puck has entered the laboratory, and the 'human mortals' will never be the same.

everywhere there is writing, which is everything we touch. The purpose of the book that you are now holding has been to provide a helical structure out of which to generate the new memes necessary for a posthuman future: new images, new questions, new concepts, and new names that can reproduce themselves like the proteins of our arguments, and the more unpredictably the better.

This is the true beginning of our end

All citations to Shakespeare's *A Midsummer Night's Dream* are from the edition by Peter Holland (Oxford: Oxford University Press, 1994). On the theater as a mode of technology see Francis Yates, *Theatre of the World* (Chicago: University of Chicago Press, 1969); William West, *Theatres and Encyclopedias in Early Modern Europe* (Cambridge: Cambridge University Press, 2002); Henry S. Turner, *The English Renaissance Stage: Geometry, Poetics, and the Practical Spatial Arts* (Oxford: Oxford University Press, 2006); Adam Max Cohen, *Shakespeare and Technology: Dramatizing Early Modern Technological Revolutions* (New York: Palgrave, 2006); Scott Maisano, 'Shakespeare's Last Act: *The Starry Messenger* and the Galilean Book in *Cymbeline*,' *Configurations* 12 (2004): 401–43; and Maisano, 'Infinite Gesture: Automata and the Emotions in Descartes and Shakespeare' in *Genesis Redux: Essays on the History and Philosophy of Artificial Life*, ed. Jessica Riskin (Chicago: University of Chicago Press, 2007), 63–84. Broader studies of early modern literature and technology include *The Renaissance Computer: Knowledge Technology in the First Age of Print*, ed. Neil Rhodes and Jonathan Sawday (London: Routledge, 2000), with focus on the printed book, and Jessica Wolfe, *Humanism, Machinery, and Renaissance Literature* (Cambridge: Cambridge University Press, 2004). John Donne's comment about the 'new Philosophy' can be found in his poem 'An Anatomy of the World: The First

Then read the names of the actors; and so grow to a point

James D. Watson's *The Double Helix* (London: Weidenfeld & Nicolson, 1968) includes a facsimile of his letter to Delbrück; Robert C. Olby's *The Path to the Double Helix* (Seattle: University of Washington Press, 1974) provides a more conventional history of the discovery, citing N. W. Timoféëff-Ressovsky, 'The Experimental Production of Mutations, *Biological Review*, 9, (1934): 411–57 on 232. Dr George Johnson is Professor Emeritus of Biology at Washington University and of Genetics at Washington University School of Medicine; from his 'On Science' column for the *St Louis Post-Dispatch*, I have cited 'Altering ANDi: Genetic engineering gets a (small) step closer to humans,' (26 January 2001; www.txtwriter.com/Onscience/Articles/alteringANDI.html). I have used W. Hamilton Fyfe's translation of Aristotle's *Poetics* for the Loeb Classical Library (New York: G. P. Putnam's Sons, 1932) and Paul Shorey's translation and edition of Plato's *Republic* for the Loeb Classical Library, 2 vols (Cambridge, MA: Harvard University Press, 1937). Stephen Halliwell's *The Aesthetics of Mimesis* (Princeton: Princeton University Press, 2002) provides a detailed discussion of the term in Plato and Aristotle; see also Erich Auerbach's *Mimesis*, trans. Willard R. Trask (Princeton: Princeton University Press, 1953); Gunther Gebauer and Christoph Wulf, *Mimesis: Culture, Art, Society*, trans. Don Reneau (Berkeley: University of California Press, 1995); Arne Melberg, *Theories of Mimesis* (New York: Cambridge University

Anniversary' (l. 205), *The Poems of John Donne*, ed. Herbert J. C. Grierson, 2 vols (Oxford: Clarendon Press, 1912). There is much scholarly work on the performative conventions of Shakespeare's theater, but I especially recommend three older works: Alan Dessen, *Elizabethan Drama and the Viewer's Eye* (Chapel Hill: University of North Carolina Press, 1977), Dessen, *Elizabethan Stage Directions and Modern Interpreters* (Cambridge: Cambridge University Press, 1984), and Robert Weimann, *Shakespeare and the Popular Tradition of the Theater: Studies in the Social Dimension of Dramatic Form and Function*, ed. Robert Schwarz (Baltimore: The Johns Hopkins University Press, 1967). I have used Paul Shorey's translation and edition of Plato's *Republic* for the Loeb Classical Library, 2 vols (Cambridge, MA: Harvard University Press, 1937) and Geoffrey Shepherd's edition of *An Apology for Poetry or The Defence of Poesy* (London: Thomas Nelson and Sons, 1965). The major antitheatrical treatises of the Elizabethan period are John Northbrooke, *A Treatise wherein Dicing, Dauncing, Vaine playes or Enterluds . . . are reproued by the authoritie of the word of God* (London, 1579); Stephen Gosson, *The Schoole of Abuse* (London, 1579) and *Playes Confuted in fiue Actions* (London, 1582); Phillip Stubbes, *The Anatomie of Abuses* (London, 1583); and John Rainolds's *Th'overthrow of Stage-Playes* (London, 1599). Important studies of the antitheatrical tradition include Jonas Barish, *The Antitheatrical Prejudice* (Berkeley: University of California Press, 1981); Jean Howard, *The Stage and Social Struggle in Early Modern England* (New York: Routledge, 1994); and Stephen Orgel, 'Nobody's Perfect: Or, Why Did the English Stage Take Boys for Women?,' *South Atlantic Quarterly*, 88, (1989): 7–29 (from which many of my quotations have

Press, 1995); Matthew Potolsky, *Mimesis* (New York: Routledge, 2006); Jacques Derrida, 'The Double Session' in *Dissemination*, trans. Barbara Johnson (London: Athlone Press, 1981), 173–285 and Derrida, 'White Mythology' in *Margins of Philosophy*, trans. Alan Bass (Hemel Hempstead: Harvester Press, 1982), 207–71; and Gérard Genette's *Mimologics*, trans. Thaïs E. Morgan (Lincoln: University of Nebraska Press, 1995), an extended essay on Plato's *Cratylus* and its legacy. I have used H. N. Fowler's translation of the *Cratylus* for the Loeb Classical Library (Cambridge: MA: Harvard University Press, 1939). I have used Jill E. Levenson's edition of *Romeo and Juliet* (Oxford: Oxford University Press, 2000). The discovery of the 'Pyramus and Thisbe' genes is described in Angelike Stathopoulos et al., 'Pyramus and Thisbe: FGF Genes that Pattern the Mesoderm of Drosophila Embryos,' *Genes and Development*, 18, (2004): 687–99 (http://www.genesdev.org/cgi/content/abstract/18/6/687). Nancy Ekardt's article 'A Rose by Any Other Name?' *The Plant Cell*, 14, (2002): 2315–17 (http://www.pubmedcentral.nih.gov/articlerender.fcgi?artid=543212) discusses Inna Guterman et al., 'Rose Scent: Genomics Approach to Discovering Novel Floral Fragrance-Related Genes, *The Plant Cell*, 14, (2002): 2325–38 (http://www.pubmedcentral.nih.gov/articlerender.fcgi?artid=151220). Shakespeare's Sonnet 130 is in Stephen Booth, ed., *Shakespeare's Sonnets* (Yale: Yale University Press). News of Oberon's potion was reported in the 14 February 2002 issue of *Nature* (http://www.nature.com/nature/journal/v415/n6873/full/415724a.html) and picked up by CNN (http://edition.cnn.com/2002/WORLD/europe/02/13/rsc.chemistry/). Quotations of Bacon are from *The Advancement of Learning* (New York: Modern Library, 2001) and from *The New Organon*, ed. Fulton H. Anderson

been taken). The best studies of Shakespeare and Ovid are Leonard Barkin, *The Gods Made Flesh: Metamorphosis & the Pursuit of Paganism* (New Haven: Yale University Press), esp. 251–70; Jonathan Bate, *Shakespeare and Ovid* (Oxford: Clarendon Press, 1993), esp. 130–44; *Shakespeare's Ovid: The Metamorphoses in the Plays and Poems*, ed. A. B. Taylor (Cambridge: Cambridge University Press, 2000); Lynn Enterline, *The Rhetoric of the Body from Ovid to Shakespeare* (Cambridge: Cambridge University Press, 2000); and Heather James, 'Ovid and the Question of Politics in Early Modern England,' *ELH*, 70, (2003): 343–73. I have used P. G. Walsh's translation of Apuleius's *The Golden Ass* (New York: Oxford University Press, 1999), which Shakespeare would have known in William Adlington's English translation as *The xi Bookes of the Golden Asse* (1566). Jean-Pierre Vernant's essay 'The Reason of Myth' may be found in his *Myth and Society in Ancient Greece*, trans. Janet Lloyd (New York: Zone Books, 1990), 203–60. A representative selection of scholarship on Shakespeare's play may be found in *A Midsummer Night's Dream: Critical Essays*, ed. Dorothea Kehler (New York: Garland Publishing, Inc., 1998); other noteworthy studies include C. L. Barber's *Shakespeare's Festive Comedy* (Princeton: Princeton University Press, 1957), 119–62; Jan Kott's *Shakespeare Our Contemporary*, trans. Boleslaw Taborski (Garden City, NY: Doubleday, 1964); and Louis Montrose's '"Shaping Fantasies": Figurations of Gender and Power in Elizabethan Culture,' *Representations*, 2, (1983): 61–94. I also recommend James L. Calderwood's book-length essay, *A Midsummer Night's Dream*, for the Twayne's New Critical Introductions to Shakespeare series (New York: Twayne Publishers, [1992]), one of the most creative and wide-ranging

(Indianapolis: Bobbs-Merrill Co., 1960); of Galileo from *The Discoveries and Opinions of Galileo*, trans. and ed. by Stillman Drake (New York: Anchor Books, 1957). Hugh of St Victor's comments on art and nature may be found in *The Didascalicon of Hugh of St. Victor*, trans. Jerome Taylor (New York: Columbia University Press, 1991), cited and discussed by William R. Newman, 'Technology and Alchemical Debate in the Late Middle Ages,' *Isis*, 80, (1989): 423–45. On 'invention' see Henry S. Turner, *The English Renaissance Stage: Geometry, Poetics, and the Practical Spatial Arts* (Oxford: Oxford University Press, 2006), 82–113. Karin Knorr-Cetina's *Epistemic Cultures: How the Sciences Make Knowledge* (Cambridge, MA: Harvard University Press, 1999) develops many of the insights of Bruno Latour, perhaps the most significant figure in the sociology of science: I especially recommend Bruno Latour and Stephen Woolgar, *Laboratory Life: The Construction of Scientific Facts*, 2nd edn (Princeton: Princeton University Press, 1986); *Science in Action: How to Follow Scientists and Engineers Through Society* (Cambridge, MA: Harvard University Press, 1987); 'Drawing Things Together' in *Representation in Scientific Practice*, ed. Michael Lynch and Steve Woolgar (Cambridge, MA: The MIT Press, 1990), 19–68; *Pandora's Hope: Essays on the Reality of Science Studies* (Cambridge, MA: Harvard University Press, 1999); and the utterly original *Aramis, or The Love of Technology*, trans. Catherine Porter (Cambridge, MA: Harvard University Press, 1996). Evelyn Fox Keller's *Refiguring Life* (New York: Columbia University Press, 1995), *The Century of the Gene* (Cambridge: Harvard University Press, 2000), and *Making Sense of Life* (Cambridge, MA: Harvard University Press, 2002) offer some of the best accounts of the history of the gene as idea and metaphor in modern biology; see also Peter J. Beurton et al.,

discussions of the play. On the dream-like qualities of *Dream* see Marjorie Garber, *Dream in Shakespeare: From Metaphor to Metamorphosis* (New Haven: Yale University Press, 1974), 59–87; on sources for and Shakespeare's characterization of Theseus see D'Orsay W. Pearson, '"Unkinde" Theseus,' *ELR*, 4, (1974): 276–98; Peter Holland, 'Theseus' Shadows in *A Midsummer Night's Dream*,' *Shakespeare Survey*, 47, (1994): 139–51; and Chapter 3 of Laurie McGuire, *Shakespeare's Names* (Oxford: Oxford University Press, 2007), on the figure of Helen in *Dream* (and in early modern literary culture as a whole), Theseus's relation to her, and on the problem of 'naming' in the play. An English translation of Marsilio Ficino's fascinating *De Vita* is available in an excellent edition as *Three Books on Life* from the MRTS series, ed. Carol V. Kaske and John R. Clark (Binghamton: Center for Medieval and Early Renaissance Studies, 1989). Henry Cornelius Agrippa's *Three Books of Occult Philosophy* is available in a modern edition with annotations by Donald Tyson to the 1651 English translation by James Freake (St Paul, MN: Llewellyn Publications, 1993). Readers interested in the occult sciences during the early modern period may wish to consult the many works of John Dee, especially his famous *Mathematical Preface to the Elements of Euclid of Megara*, first published 1570 (New York: History of Science Publications, 1975), and Giambattista della Porta's *Magia Naturalis*, first published in Naples in 1558 and translated into English anonymously in 1658 as *Natural Magick* (and available in a facsimile from Basic Books, Inc, 1957). Good historical introductions to the occult in the early modern period are Francis Yates, *Giordano Bruno and the Hermetic Tradition* (Chicago: University of Chicago Press, 1964); D. P. Walker, *Spiritual and Demonic*

eds., *The Concept of the Gene in Development and Evolution* (Cambridge: Cambridge University Press, 2000). Darwin's theory of 'pangenesis' and his letter to George Romanes are discussed by Donald R. Forsdyke, 'The Origin of Species, Revisited: A Victorian who Anticipated Modern Developments in Darwin's Theory,' *Queen's Quarterly*, 106, (1999): 112–33 (http://post.queensu.ca/~forsdyke/evolutio.htm). Paul Feyerabend's *Against Method*, rev. ed. (London and New York: Verso, 1988) is a classic in the philosophy of science. Francis Crick's memoir, *What Mad Pursuit* (New York: Basic Books, 1988) has recently been supplemented by Matthew Ridley's biography, *Francis Crick: Discoverer of the Genetic Code* (New York: Harper Collins, 2006). On experimentation as performance see Robert Crease, *The Play of Nature* (Bloomington: Indiana University Press, 1993); on *mimesis* in anthropology see Michael T. Taussig, *Mimesis and Alterity: A Particular History of the Senses* (New York: Routledge, 1993). I have cited John Keats's 'Ode on a Grecian Urn' from *The Complete Poems*, ed. John Barnard (New York: Penguin, 1988). On concepts of code and information in molecular biology see Lily Kay, *Who Wrote the Book of Life?* (Stanford: Stanford University Press, 2000); Georges Canguilhem, 'Epistemology of Biology' and 'Knowledge and the Living,' both in *A Vital Rationalist*, ed. Francois Delaporte and trans. Arthur Goldhammer (New York: Zone Books, 1994), 67–90 and 287–319; and Jean Baudrillard's *Simulations*, trans. Paul Foss, Paul Patton, and Philip Beitchman (New York: Semiotext(e), 1983). Wendy Doniger's 'Sex and the Mythological Clone' may be found in *Clones and Clones*, ed. Martha Nussbaum and Cass R. Sunstein (New York: Norton, 1998), 114–38. Among much excellent scholarship on the printed Shakespearean text, I recommend Margreta De Grazia

Magic from Ficino to Campanella (London: The Warburg Institute and the University of London, 1958/Nendeln: Kraus Reprint, 1969); William Eamon, *Science and the Secrets of Nature: Books of Nature in Medieval and Early Modern Culture* (Princeton: Princeton University Press, 1994); Pamela Smith, *The Business of Alchemy: Science and Culture in the Holy Roman Empire* (Princeton: Princeton University Press, 1994); and William Newman, *Promethean Ambitions: Alchemy and the Quest to Perfect Nature* (Chicago: University of Chicago Press, 2004). Elizabeth Sewell's *The Orphic Voice: Poetry and Natural History* (New Haven: Yale University Press) is a beautiful and entirely original study that attempts to think beyond the divide between 'poetry' and 'science.' The best study of 'judgment' in the classical and early modern periods is David Summers, *The Judgment of Sense: Renaissance Naturalism and the Rise of Aesthetics* (Cambridge: Cambridge University Press, 1987). On Bottom and the 'rude mechanicals' see especially Robert Weimann, *Author's Pen and Actor's Voice*, ed. Helen Higbee and William West (Cambridge: Cambridge University Press, 2000), and Patricia Parker, *Shakespeare from the Margins* (Chicago: University of Chicago Press, 1996), 83–115. Adventurous readers interested in the notion of 'becoming-animal' will want to acquaint themselves with Gilles Deleuze and Félix Guattari's mind-bending *A Thousand Plateaus: Capitalism and Schizophrenia*, trans. Brian Massumi (Minneapolis: University of Minnesota Press, 1987). Giorgio Agamben's *The Open: Man and Animal*, trans. Kevin Attell (Stanford: Stanford University Press, 2004) is especially relevant to a study of the status of animals in philosophy; critics of Shakespeare are beginning to show greater interest in early modern animals, notably Erica

and Peter Stallybrass, 'The Materiality of the Shakespearean Text,' *Shakespeare Quarterly*, 44, (1993): 255–83, and the work of Random Cloud-McCleod, especially "The very names of the Persons': Editing and the Invention of Dramatick Character' in *Staging the Renaissance,* ed. David Scott Kastan and Peter Stallybrass (New York: Routledge, 1991), 88–96. On the 'posthuman' see the special issue of *Configurations*, 10, (2002) edited by Timothy Lenoir, esp. Lenoir's 'Makeover: Writing the Body into the Posthuman Technoscape' (203–20), citing Rodney Brooks, *Flesh and Machines: How Robots Will Change Us* (New York: Pantheon, 2002); N. Katherine Hayles, *How We Became Posthuman* (Chicago: University of Chicago Press, 1999); Hayles, *My Mother Was a Computer* (Chicago: University of Chicago Press, 2005); Hayles, 'Refiguring the Posthuman,' *Comparative Literature Studies*, 41, (2004): 311–16; Richard Doyle, *On Beyond Living: Rhetorical Transformations of the Life Sciences* (Stanford: Stanford University Press, 1997); Doyle, *Wetwares: Experiments in Postvital Living* (Minneapolis: University of Minnesota Press, 2003); Doyle, 'Emergent Power: Vitality and Theology in Artificial Life,' in Lenoir, ed. *Inscribing Science*, 304–27; and the engaging study of computer simulations, robots, and artificial life by Steven Levy, *Artificial Life: The Quest for a New Creation* (New York: Pantheon Books, 1992): Chapter 2 of Bruce Clarke's forthcoming *Posthuman Metamorphosis* (New York: Fordham University Press) examines systems theory in relation to *A Midsummer Night's Dream*. Details about the Riken Corporation's miraculous DNA Book may be found in Jun Kawai and Yoshihide Hayashizaki, 'DNA Book,' *Genome Research*, 13, (2003): 1488–95 (http://www.genome.org/cgi/content/full/13/6b/1488) and at http://genome.gsc.riken.go.jp/DNA-Book. John Milton's

Fudge, *Perceiving Animals: Humans and Beasts in Early Modern English Culture* (New York: St. Martin's, 2000); Fudge, *Brutal Reasoning: Animals, Rationality, and Humanity in Early Modern England* (Ithaca, NY: Cornell University Press, 2006); Fudge, *Animal* (London: Reaktion Books, 2002); Bruce Boehrer, *Shakespeare Among the Animals: Nature and Society in the Drama of Early Modern England* (New York: Palgrave, 2002), with suggestions for further reading; Boehrer, 'Economies of Desire in *A Midsummer Night's Dream*,' *Shakespeare Studies*, 32, (2004): 99–117; and Laurie Shannon, *Zoographies of Knowledge: Animal Agency and the Early Modern Constitution* (forthcoming). On nanotechnology and literature see Colin Milburn, 'Nanotechnology in the Age of Posthuman Engineering: Science Fiction as Science,' *Configurations*, 10, (2002): 261–95, and Milburn, 'Nano/ Splatter: Disintegrating the Postbiological Body,' *New Literary History*, 36, (2005): 283–311. I have used W. Hamilton Fyfe's translation of Aristotle's *Poetics* for the Loeb Classical Library (New York: G. P. Putnam's Sons, 1932); John Henry Freese's translation of the *Rhetoric*, also for the Loeb (Cambridge, MA: Harvard University Press, 1991); and a modern facsimile of George Puttenham's *The Arte of English Poesie*, ed. Edward Arber (n.p: A. Constable and Co. Ltd., 1906 / reprint. [Kent, OH]: Kent State University Press, 1970). Donna Haraway's classic 'A Manifesto for Cyborgs' may conveniently be found in *The Haraway Reader* (New York: Routledge, 2004), 7–45. Murray Krieger's comments on Renaissance poetic language may be found in his 'Presentation and Representation in the Renaissance Lyric: The Net of Words and the Escape of the Gods,' in *Mimesis, from Mirror to Method, Augustine to Descartes*, ed. John D. Lyons and Stephen G. Nichols, Jr.

Areopagitica is conveniently collected in *Complete Poems and Major Prose*, ed. Merritt Hughes (New York: Macmillan, 1957). For discussion of species definitions and chimeric substances see the special issue of the *American Journal of Bioethics*, 3.3, (Summer 2003), with a target article by Jason Scott Robert and Françoise Baylis followed by 21 responses. Richard Lewontin's *Biology as Ideology: the Doctrine of DNA* (New York: Harper Perennial, 1992) is an accessible introduction to the 'myth' of DNA. Further discussion of chimeras, by turns enlightening, amusing, and infuriating, may be found in the transcripts of the President's Council on Bioethics, especially the 25 September 2003 meeting (www.bioethics.gov/transcripts/sep03/sep5ull.html), the 16 October 2003 meeting (www.bioethics.gov/transcripts/oct03/oct16full.html), and the 4 March 2005 meeting (www.bioethics.gov/transcripts/march05/march04full.html). References to Shakespeare's *The Winter's Tale* are from J. H. P. Pafford's Arden edition (New York: Methuen, 1963). I have used Geoffrey Shepherd's edition of *An Apology for Poetry or The Defence of Poesy* (London: Thomas Nelson and Sons, 1965). Examples of the new zoology of biotechnology may be found in Robert and Baylis; in Michael J. Reiss and Roger Straughan's *Improving Nature?: The Science and Ethics of Genetic Engineering* (Cambridge: Cambridge University Press, 1996), and at many online news outlets. The work of Donna Haraway, with representative essays collected in *The Haraway Reader* (New York: Routledge, 2004), has been especially influential on critical interest in the status of the animal, as has Jacques Derrida's 'The Animal That Therefore I Am (More to Follow),' trans. David Wills, *Critical Inquiry*, 28, (2002): 369–418, as exemplified by *Zoontologies: The Question of the Animal,* ed. Cary Wolfe (Minneapolis:

(Hanover, NH: University Press of New England, 1982), 110–31. Bruno Latour's *We Have Never Been Modern*, trans. Catherine Porter (Cambridge, MA: Harvard University Press, 1993) is his most influential, schematic, and polemical book, but Latour's arguments are always best approached through his detailed case studies, for which readers should read the right-hand side of this book.

University of Minnesota, 2003). The notion of the meme appears in Richard Dawkins, *The Selfish Gene*, 2nd edn (Oxford: Oxford University Press, 1989), and a few authors are more 'memetic' than Shakespeare.

Index

actor 8, 28, 30, 43, 48, 52, 70, 72, 73, 76, 80, 81, 82, 85, 93, 102
Advancement of Learning (Bacon) 29, 103
Agamben, Giorgio 88, 100
Agrippa, Henry Cornelius 40, 42, 44, 52, 54, 62, 66
alchemy 42. *See also* magic; occult
animal 6, 9, 14, 15, 58, 68, 78, 80, 83, 88, 90, 96, 100, 101
Apuleius 22, 56, 57
  The Golden Ass 12, 57, 60, 66
Aristotle 7, 35, 37, 47, 49, 94
  Physics 37
  Poetics 7, 15, 37, 47, 49
art 13, 15, 39, 82. *See also techne*
  and nature 35, 37, 42, 49, 56, 64, 84, 86, 88, 91, 92
Arte of English Poesie (Puttenham) 68, 84
astrology 40, 42
Athens 6, 19, 20, 26, 30, 32, 54, 59, 79, 106
atom 3, 41, 43, 90
Avery, Oswald 55

Bacon, Francis 29, 35, 49, 95, 105
  Advancement of Learning 29, 103
Barthes, Roland 107
Baylis, Françoise 87, 89
Beadle, George 63
Berlin 3, 5
Bible, the 76
  Scripture 33, 76
bioethics 89, 101
biotechnology 2, 7, 35, 61, 93, 101, 108
body, human 45, 46, 48, 50, 52, 622, 64, 76, 81, 92, 98, 100, 102, 107
Boehrer, Bruce 104
Bottom 23, 26, 66, 68, 70, 72, 79, 81, 82, 85
Brooks, Rodney 85
Burbage, James 82
Burbage, Richard 82

Cambridge 5, 55, 57, 71
cell 24, 45, 52, 57, 63, 75, 77, 81, 107. *See also* stem cell
Chaucer 22
chimera 84, 89, 95, 99, 100, 101
Church of England 8, 106
cloning 5, 71, 81, 87, 99, 100, 101
comedy 6, 22, 66, 82, 92
comets 12, 14, 31, 40, 44, 56, 93

125

computer 47, 69, 71, 77, 81, 83
computer code 5, 61, 77, 83
computer program 5, 61, 83
Crease, Robert 47
Crick, Francis 3, 47, 53, 55, 57, 59, 61, 63, 65, 67, 71, 75
Cupid 60
cybernetics 81, 85

Darwin, Charles 43, 45
Dawkins, Richard 67, 69, 71, 105
Dee, John 40
Defence of Poesy (Sidney) 8, 20, 54, 93
Delbrück, Max 3, 5
Demetrius 26, 56, 62, 64, 60, 68, 94, 96, 98
Derrida, Jacques 103
Descartes, René 50, 83
DNA 2, 3, 5, 23, 53, 55, 57, 61, 63, 73, 75, 77, 87, 89, 101, 105
DNA Book Series 87
Doniger, Wendy 71
Donne, John 4
Donohue, Jerry 53
Doyle, Richard 81, 83, 85
dramatology 4, 17, 37, 78

Eamon, William 42
Egeus 26, 58, 60
Ekardt, Nancy 23, 27
Empedocles 7
eroticism 30, 66, 74, 88
ethics 15, 28, 101. *See also* bioethics
evolution 3, 45, 107

Feyerabend, Paul 45, 49
Ficino, Marsilio 40, 42, 44, 46, 52, 56, 58
fiction 7, 18, 33, 35, 37, 41, 47, 83, 95, 97, 101, 102
Franklin, Rosalind 55
Freud, Sigmund 66, 68
Fudge, Erica 74, 104

Galileo 29, 31, 33, 35, 51, 63, 95
Gardner, Martin 86
gene 5, 21, 23, 27, 43, 57, 65, 73, 75, 77, 79, 99, 105, 107
genetic code 61, 63, 71
genetic engineering 5, 7, 35, 61. *See also* biotechnology
genetic mutation 3
genetic transmission 67, 76
genome 75, 77, 79, 81
  Human Genome Project 79
Globe Theatre 4
God 14, 54, 74, 97
Golden Ass, The (Apuleius) 12, 57, 60, 66
Gosson, Stephen 8
Grassi, Horatio (pseudonym Sarsi) 31

Halliwell, Stephen 9
Helena 62, 64, 66, 68, 92, 94, 96, 98, 100, 102
helix 2, 53, 55, 57, 86, 96
Hermes 52
Hermia 26, 60, 62, 64, 66, 68, 72, 92, 96, 98, 100
Heywood, Thomas 8

Hippolyta 22, 24, 26, 30, 32
Hugh of St. Victor 37
Human Genome Project. *See* genome
Humphrey, N. K. 107
hybrid 14, 22, 23, 50, 84, 88, 95, 102, 104, 106

imitation 4, 6, 13, 15, 35, 37, 42, 44, 95, 97, 102, 107. *See also mimesis*
invention 29, 37, 93, 106

Johnson, George 5, 61
Jonson, Ben 8, 82

Kay, Lily 63
Keats, John 59
Keller, Evelyn Fox 43, 63, 75
King Lear (Shakespeare) 8
Knorr-Cetina, Karin 39, 43, 47
Kott, Jan 30
Krieger, Murray 106, 108

language 4, 12, 16, 48, 50, 52, 64, 76, 104
  and Galileo 29, 31, 33
  mythic 16, 18
  of poetry 76, 86, 89, 94, 98
  of science 2, 42, 43, 63, 65, 79, 83, 87, 89
  Shakespeare's use of 27, 108
  Socrates's philosophy of 11, 13, 15, 17, 19
Latour, Bruno 35, 41, 49, 61, 102, 104

Lenoir, Timothy 83
London 8, 10, 24, 66
Lysander 26, 53, 56, 58, 60, 62, 64, 66, 68

machine 4, 7, 14, 40, 73, 83, 85, 96
MacLeod, Colin 55
magic 4, 12, 14, 40, 42, 44, 52, 53, 54, 56, 60, 62, 66, 76, 80. *See also* alchemy, occult
Marlowe, Christopher 20
mathematics 33, 47, 63, 67, 71
McCarty, Maclyn 55
McCleod, Randall-Random Cloud 77, 79
McLoughlin, Claire 29
Measure for Measure (Shakespeare) 26
meme 107, 109
metaphor 27, 39, 46, 48, 52, 61, 63, 65, 75, 78, 80, 81, 82, 83, 84, 86, 88, 92, 93, 94, 96, 98, 104, 107
metaphormosis 92
Miller, Thomas 99
Milton, John 87
mimesis 4, 7, 9, 19, 33, 35, 37, 39, 43, 44, 47, 48, 51, 53, 55, 61, 71, 72, 78, 93, 102
Mirandola, Giovanni Francesco Pico Della 54
molecular biology 3, 43, 61, 63, 73, 83
moon 12, 24, 28, 30, 36, 38, 40, 51, 56, 58, 88

music 9, 13, 52, 54, 93
mutation 3, 5, 57, 60, 67, 69, 73, 77, 78, 79, 96
myth 12, 14, 16, 18, 20, 21, 22, 23, 26, 29, 30, 32, 34, 40, 45, 52, 60, 62, 76, 78, 107

natural selection 47, 67, 69, 71
nature 33, 34, 35, 37, 38, 39, 40, 42, 44, 49, 52, 56, 63, 64, 69, 84, 86, 88, 91, 93, 95, 97, 101, 103, 105
Newman, William 37, 42

Oberon 30, 32, 34, 38, 46, 54, 56, 59, 60, 62, 66, 86, 88, 94
occult 14, 35, 40, 42, 44, 52, 54, 58, 68, 80, 96. *See also* alchemy; magic
onomapoietics 13, 17, 21, 43
onomastikos 13, 41
organism 3, 5, 47, 52
organon 17, 49
Orlando Furioso (Ariosto) 31, 33
Orpheus 52, 54
Ovid 22

Pauling, Linus 53
philosophy 4, 7, 11, 15, 17, 18, 21, 33, 39, 42, 46, 50, 79, 88
physics 39, 43, 47, 51
Physics (Aristotle) 37
Poetics (Aristotle) 7, 15, 37, 47, 49
Plato 4, 7, 9, 18, 20, 35, 59
poesy. *See* poetry

poet 12, 20, 29, 31, 37, 41, 49, 54, 59, 60, 64, 81, 84, 91, 93, 95, 97, 100, 103, 105, 108
poetry 6, 7, 20, 25, 27, 29, 33, 35, 39, 45, 52, 55, 76, 93, 96, 97, 105
poiesis 37, 108
Polixenes 91
Pope, Alexander 73
Porta, Giovanni Battista Della 40
praxis 15
Proteus 49, 78
Prynne, William 74
Psyche 60
Puck 13, 30, 56, 78, 80, 82, 86, 108
Puttenham, George 68, 84
  Arte of English Poesie 68, 84
Pyramus 21, 23, 26, 60, 72, 79

Rainolds, John 10
Robert, Jason 87, 89
robot 79, 81, 85
Romanes, George John 45
Romeo and Juliet (Shakespeare) 19, 21, 23, 26
rose 19, 21, 23, 25, 27, 38, 92, 99
Rowe, Nicholas 73
Royal Shakespeare Company 27
Royal Society of Chemistry 27, 29

Schrödinger, Erwin 63, 75
science 2, 5, 31, 23, 34, 35, 39, 42, 43, 44, 45, 49, 58, 63, 75, 81, 89, 100, 102, 108
  and literature 7, 35

and *mimesis* 35, 42
and poetry 33, 93, 100
Scripture. *See* Bible, the
seasons 36, 38, 40, 56, 64, 92
Sell, Charles 29
Sewell, Elizabeth 48, 50
sexuality 10, 28, 32, 66
Shakespeare, William. *See also King Lear; Measure for Measure; Midsummer Night's Dream, A; Romeo and Juliet; Winter's Tale, The;* individual character's names
  and computer code 5, 61
  and *mimesis* 37, 44, 48, 78
  and myth 20, 23, 24, 26, 34, 40
  and names 17, 19, 21, 23
  and nature 2, 27, 35, 38, 59, 60, 92, 106
  and poetry 25, 27, 54, 89
  and science 7, 23, 29, 30, 36, 67, 71, 73, 81, 85, 89, 91, 101
  and theatre 4, 50, 81
Shannon, Laurie 104
Sidney, Sir Philip 8, 20, 54, 93, 95, 97, 101
Defence of Poesy 8, 20, 54, 93
Smith, Pamela 42
Socrates 4, 6, 9, 11, 13, 15, 17, 19, 20, 39, 101, 105, 106
song 58, 60, 76, 98
sonnet 25, 62
species 9, 11, 21, 23, 25, 29, 45, 49, 74, 84, 86, 89, 92, 94, 100, 101, 102

stars 12, 14, 31, 40, 44, 56, 93. *See also* sunspots, comets
stem cell 99, 100. *See also* cell
Stubbes, Philip 10
sun 12, 25, 28, 30, 51, 56
sunspots 31
symbol 6, 14, 16, 30, 34, 42, 57, 61, 80

Taussig, Michael 51, 53
techne 15
technology 4, 5, 37, 39, 55, 73, 77, 79. *See also* biotechnology
theatre 4, 6, 7, 10, 14, 34, 37, 45, 48, 50, 52, 58, 72, 73, 77, 81, 82, 100, 106
Theseus 17, 19, 23, 24, 26, 28, 30, 32, 41, 45, 59, 64, 66, 83, 104, 106, 108
Thisbe 21, 23, 26, 60, 66, 72, 75
Timoféëff-Ressovsky, T. M. 3, 5
Titania 27, 29, 30, 32, 34, 38, 46, 48, 50, 52, 56, 62, 66, 68, 70, 72, 86, 88, 94, 99

Uexküll, Jakob von 90

Vernant, Jean-Pierre 14, 16

Watson, James 3, 53, 55, 57, 63, 79
Wegenstein, Bernadette 79
Winter's Tale, The (Shakespeare) 91

xenotransplantation 99, 100